napa bulletin

Electronic Technologies and Instruction: Tools, Users, and Power

■ Frank A. Dubinskas and James H. McDonald, eds.

National Association for the Practice of Anthropology
A Unit of the American Anthropological Association

NAPA Bulletins are occasional publications of the National Association for the Practice of Anthropology, a Unit of the American Anthropological Association.

Frank A. Dubinskas and James H. McDonald
General Editors

Library of Congress Cataloging-in-Publication Data

Electronic technologies and instruction : tools, users, and power / Frank A.
  Dubinskas and James H. McDonald, eds.
    p.   cm.—(NAPA bulletin : 12)
  Includes bibliographical references.
  ISBN 0-913167-57-6
    1. Anthropology—Study and teaching—Data processing.
  2. Anthropology—Computer-assisted instruction. I. Dubinskas, Frank A.
  (Frank Anthony) II. McDonald, James H. III. Series.
  GN42.5E44   1993
  301′.0285—dc20   93-27139
                                                        CIP

ISBN 0-913167-57-6

Contents

Introduction

Knowledge Building and Knowledge Access: Teaching with Electronic Tools

Frank A. Dubinskas

This volume, *Electronic Technologies and Instruction: Tools, Users, and Power,* addresses issues at the heart of our changing anthropological teaching and its politics—the processes of building knowledge and creating access to knowledge resources. The changes we discuss emerge from and are built with new electronic teaching technologies. The introduction of these technologies transforms the media, speed, density, and the character of the work of students and teachers, and it alters the organization and presentation of information to both. Some of these technologies also provide information on the learning process itself by automatically recording data on accuracy, quality, and frequency of use.

The new teaching tools, however, are not simply dumb instructions with an "impact" on teaching; but, rather, they are active ingredients in a process of redistributing knowledge and power among students, teachers, administrators, and the organizations or communities where they are used. As Stephen Barley's ethnographies (1986, 1990) of medical CAT scanners demonstrate, introducing a high-tech tool creates an *occasion* for restructuring social relations, but tools by themselves do not determine social and organizational—or pedagogic—outcomes. This inability to predict social consequences from the technology alone necessitates a detailed social study of implementation contexts to help understand the dynamics of the process. At the same time, the character of a technical system does have an *influence* in shaping the possibilities for its use. There is a flexible balance between the relative influence of tools *versus* users, and because electronic tools are often as flexible as their human users, this balance is a continually shifting one.

Our aim in this volume is to shed light on this complex and shifting balance by presenting five studies of new tools and teaching processes in their social contexts. Among the technical systems we address are:

- Interactive tutorial software (Hamill and Marchant)
- Hypertext software for communication and teaching (Mason Weiss, Metzger, and McDonald; Bader and Nyce)
- Telecourses (Segal)

- Computers and teleconferencing (McDonald; Mason Weiss, Metzger, and McDonald)

The social contexts range from an individual user to classroom settings to dispersed electronic networks of students ("distance learning") to community and regional systems of educational administration.

As we examine these new tools and learning contexts, the concept of "knowledge management," comprising knowledge building and access to knowledge, can help separate different issues and realms of analysis. The teaching and learning processes are centrally concerned with building and spreading knowledge. These two are intimately linked, because the classroom learning process builds knowledge; and those same classroom interactions open access to knowledge bases. Differentiating building from access, however, creates an order to how we address the issues. The first aspect, knowledge *building* focuses on the ways that students learn to think about and perform in a subject area. Here we explore the notions of how well (or how fast or how accurately) students become familiar with either a body of knowledge—a "knowledge base"—or a set of intellectual tools that allow them to create and navigate through such knowledge. In other words, "Did they learn the stuff?" Knowledge *access* opens a related discourse on the politics of how students might accomplish that learning. Here we examine social barriers to and facilitators of learning, especially those rooted in the inherent power asymmetry between teachers (or educational institutions) and their students. The discussion of access also widens the scope of discourse to contexts outside the classroom that impinge on the learning process, such as schools, communities of students and teachers, administrators, and social, political forces, and agents from ever widening arenas of relevance. Our articles consider contexts as narrow as learning alone or in small groups, to the broad questions of political and cultural hegemony in the United States.

New electronic technologies, in every case, have a central role in stimulating a transformation of the teaching and learning experience. Two issues become central to this transformation: (1) the user's encounter with new technical systems and (2) the transformations of power relations that accompany or are engendered by these tools. By framing these issues as knowledge building and knowledge access, we propose a model for examining similar new technology implementations that are not strictly focused on teaching or on anthropology. In the following section, I define and describe several dimensions for examining our five case studies—tool transparency; task clarity; control, flexibility, and skill; and empowerment. These dimensions also recast our teaching studies more broadly as examples of new technology implementation.

"Glass Tools" and Knowledge Building

Tool transparency addresses the character of electronic tools and the intimate relation between tasks and tools. Our case studies focus primarily on the user's point of view rather than the tool designer's, because most of the tools we discuss appear to the naive student user as "finished" objects. However, it is also necessary to consider the high potential flexibility of software tools and, hence, their malleability in the hands of users. Further, this malleability leads us to consider the tool *design* process, too, because sophisticated users become de facto "designers" as they learn about and modify their flexible tools. Considering the tool design process also reminds us that the user-to-designer relationship is critical for choosing or producing effective tools. The tool transparency issue directs our attention to whether and how a tool facilitates (or impedes) the user in getting a task done (Ehn 1988; Norman 1988; Seely-Brown and Duguid 1990).

The "glass tool" or "glass box" tool is a metaphoric extension of the engineer's notion of a "black box." While a "black box" is an opaque technology that masks the process or reasoning by which it functions, originally a "glass box" meant a tool that allows the user to see the underlying processes that make the tool work. At first blush, this seems like a positive attribute; however, the problem of 'transparency' is more complex. Is a tool "transparent" enough to the user so that the user can focus on the *task* to be accomplished, not on the details of how to use the tool itself? Or rather, does the tool embody knowledge about the task that must be made "visible" in order for the user to understand the task process? This tool-to-task relationship links four issues with questions we can ask of the cases described in this volume.

1. Tool transparency: Do you *see* or see *through* the tool, and how hard is the tool to use?

2. Task clarity: Are the tasks clear to students (or other users), and is the relation of the tool to the task also clear?

3. Control: Does the user have control of the tool, or does the tool force the user into a predetermined mode of work?

4. Flexibility and skill: If a tool is flexible, who is empowered by skills or knowledge to take advantage of this flexibility?

Tool transparency is concerned with what a tool reveals or conceals, and thus also how easy a tool is to use. No matter how 'flashy', sexy, or engaging a tool may seem on its surface, getting to the task objective should be the prime driver for picking a tool. Some tools force you to jump through hoops to get your work done, and other tools ease the path, enhancing your ability to do a good job. The first kind of "hoops" tool requires a lot of learning and attention to the tool-manipulation process itself, but manipulating the tool does not enhance your knowledge or productivity in the domain of your objective. Take word processors as an example. If you had to type the source code for every formatting action you wished to execute in a word-processing program, you would either produce less elegant, less legible documents, or you would spend hours adding cryptic instructions in brackets like "{.ital}" before or after *every* change of style-like italicization, indentation, spacing, or alignment. You must learn a new vocabulary and syntax for creating style—a "command language"—but this new learning task only marginally facilitates your primary writing objective. The system is opaque and must be deciphered in order to get to the task of writing. By contrast, a menu or icon-driven word-processing system that is also "WYSIWYG"[1] facilitates producing text the way you want it to look. The complexity of manipulating text on the screen or page is hidden in the software's source code behind graphic icons or menu items that are readily accessible. This system makes formatting a document relatively easier and allows the user to focus on the fundamental objective of writing.

In this case, the word-processing technology mediates between the user and the task; so it makes sense for its functions to be relatively invisible or transparent. You see through the word processor to the content of your document, but tool transparency is not always the most desirable aim. Some software, including some of the educational systems we describe here, embodies aspects of the subject matter in the way it works. In this case, you *want* to make the internal processes or manipulations of the technology visible. For example, in software that sorts, correlates, or analyzes data, it may be a great advantage to have these processes visible or at least accessible to students, so they can learn the analytic procedures of the system. A "glass tool" allows the student to see the inner workings of an analytical process—learning to reason, rather than just focusing on results.

The value of opacity or transparency depends on context. One might want a tool to provide different levels of visibility of its logic for different users, based upon their expertise. Beginning and advanced students might use the same tool to different ends, so the capability to alter the tool itself is an important criterion to consider. Seely-Brown and Duguid (1990) propose three kinds of transparency:

- Domain transparency
- Internal transparency
- Organizational transparency

Domain transparency is the ability to make the causal reasoning behind a technical system apparent to the user. But, as they point out, some systems are inherently complex or abstruse, so that simply laying them bare does not make them accessible. "But it is possible in certain circumstances to build a representative *model* of the relevant causality that allows users to develop a helpful sense of their role and their tool's role in ongoing activity" (1990:29). The icon-driven software of Apple's *Macintosh* program exemplifies such a model. It represents activities in the operating system of the computer by translating and metaphorically simplifying them into visual symbols that evoke common experiences; its trashcan icon, for example, represents the "delete" function. *Internal transparency* is a less mediated form of providing access to logic. Here, the details of "hidden reasoning" are presented in barer form to the user, so a process can be traced throughout its execution. The goal of this revelation is not just exposure (e.g., for learning), but to allow intervention in the system. This kind of modifiability is central to Shoshana Zuboff's argument (1988) that for knowledge to expand, technologies must allow for change based upon human judgment. Finally, *organizational transparency* is aimed at expanding the context of tool and task, away from the single machine and individual user, to consider whole social systems of technology use in their organizational, cultural, social, and political arenas. While this notion seems second nature to anthropologists, it is more novel to computer scientists. Here, for instance, we may critique the electronic "distance learning" systems of our articles based on whether they adequately create a *community* of discourse or whether they allow only dyadic interchanges between students and teachers.

Task clarity is seeing the tool-and-user relationship from the opposite side: "What is it that you really want to do?" Being clear about the task helps avoid 'techno-marvel' seduction by flashiness during tool choice. Task clarity is crucial for evaluating the utility of tools. In any attempt to implement a new technology, it is critical to assess the needs or objectives of users first. Keeping user needs in mind helps clarify whether you want to keep the task clear and the tool invisible; or whether you want to embed some of the learning task—its logic or its associated data—in the tool itself and make this visible, too. Taking the user's point of view seems like an obvious good thing, but it is by no means common or easy. It requires a serious engagement of real users in the tool design process, but most anthropologists, teachers, and industrial trainers do not have easy access to the design process. However, as tools become more flexible, we can try to bring these flexibilities out for students in real-world testing (as several of our articles illustrate). What the scholarship on user-oriented computer design, especially that from Scandinavia (e.g., Bjerknes, Ehn, and Kyng 1987; Ehn 1988), tells us is that "expert" designers cannot be trusted to take the user's point of view; they are too committed to their own logic of design. Effective systems require a social engagement of users and designers over the design process, where continuous feedback and accommodation are important for creating a user-serving system.

Control and flexibility are complementary dimensions of the tool and user system that strongly influence learning capabilities. Who controls the tool use when it is in the hands of users? Does the tool force a particular avenue or direction upon the student (or teacher), or does it leave opportunities for user discretion and judgment? Zuboff, in her extensive ethnographic study of new technology introductions (1988), gets at the heart of this issue when she distinguishes *automating* from what she calls *informating* systems. *Automating* systems are ones that "replace human effort and skill with a technology that enables the same processes to be performed at less cost and with more control and continuity" (Zuboff 1985:8). An automating system's focus is on a Taylorist separation of control away from the hands of practitioners—be they workers or students—and its transfer to the hands of managers, teachers, or computer designers, whose controlling power is now embedded in the technology. Applying this concept to teaching tools would create systems for rote learning that mask reasoning processes and diminish the possibilities for discretionary choice or experimentation by students. The electronic glass box can be made opaque, keeping both control and opportunity away from the user.

By contrast, Zuboff's concept of "informating" recognizes that new electronic technologies, even automating systems, create new information in the very process of control.[2] New information is created and can be redeployed to users on computer screens whenever some mechanical or computational activity is conducted by a computer. This new information capability consists of two kinds: one makes information about a current process visible and the other makes remote and previously hidden information visible (and accessible) through networks. If this informating capability of electronic systems is exploited positively, it can open new possibilities for learning, experimentation, and problem solving. While Zuboff is most concerned with systems for factory and office productivity, her central point is that the information-creating capabilities of new electronic systems should be used to enhance those intellective skills at which humans alone excel: creativity, learning, judgment, and problem solving. This requires putting discretionary control—the ability to make choices and to experiment—into the hands of users. As Bader and Nyce point out in this volume, what sounds in American culture like an obvious good aim—individual choice—is actually in conflict with the competing ethos of teacher-as-expert. In practice, these competing needs for control (or guidance) *versus* choice must always be balanced in the face of specific learning objectives.

Flexibility and skill are inextricably entwined in the practical relationships between tools and users. Flexibility in a tool implies that a user may manipulate the tool to reconfigure or create new knowledge or to fit the tool to an emerging learning context. The tool itself must permit this kind of access to inner workings—exhibit some transparency—for a user to modify it. Flexible capacity in a tool is moot, however, if a user lacks the skill to deploy it. Skill levels and tools must thus be matched in order to gain the most effective learning. Because tools may be flexible both to students and to teachers, teacher-oriented flexibility is as important as student-user flexibility. Teachers may have access to flexibility in tools at a deeper level than students, and this capability can let them adjust characteristics of the tool to different student constituencies. Flexible access to different levels of tool difficulty also lets teachers plan a staged entry for students into the mysteries of complex tools or algorithms without overwhelming them.

Limits on adjustment, however, also come from faculty who cannot accommodate the whole range of student abilities. Few teachers have students with uniform skills in computer use—or in anthropology, for that matter. Some compromise must be sought between tailoring tools for individuals *versus* clustering students with similar skills. The issue becomes, "Where will flexibility lie, and who controls it?" Some flexibility can be in the tool, some in the students, and some in the work a teacher does to prepare the tool for use. Creating a balance across this spectrum should be an ongoing process. A choice of tools should keep in mind that technologies will change and students will learn and increase their skills. Because the aim of the process is building knowledge, a strategy for tool selection can take into account the self-accelerating properties of some technology-rich learning processes. A software system, for instance, which models and thus makes visible some complex and otherwise abstruse domain, simultaneously, imbues its user with an intellectual and a technical skill. The intellective skill is in understanding the dynamics of a complex system, and the technical skill is an ability to extend the same tool to modeling new systems.

Knowledge Access and Empowerment

These discussions emphasize that technologies and their social contexts are interactive systems. Questions about appropriate tools must consider both the objectives and the actual practice of learning in each context. Tools can be empowering vehicles for building knowledge, but at the same time, they must be accompanied by social empowerment. The locus of control over the learning process lies not only with tools, teachers, and classrooms, but also in the structures and power of the school systems and communities where learning takes place. The question of empowerment and its relation to knowledge, learning, and the social control of the learning process are emphasized in the articles by McDonald and by Bader and Nyce.

A discussion of skills and flexibility leads easily to an expanded political discourse on knowledge access. Access is simultaneously a function of tools—their design and flexibility, the skills of users (students *and* teachers), and the social contexts in which these skills and tools are deployed. These contexts have multiple levels that affect the empowerment of users. Among the basic barriers to use are an absence of hardware or the scarcity of anything like the technology in a target population. The latter problem is illustrated in McDonald's article, where junior college students are recruited from a wide demographic spectrum and study through a remote electronic connection in a 'computer conference'. In his experience, the least computer-experienced students rapidly drop out in the face of learning a new communications tool (computer conferencing software) as well as the unfamiliar subject matter of their course. The tool's opacity seems to be a critical barrier for marginal students, and it is the poorer, older, female, and "nontraditional" junior college student who falls out the soonest, making "technology familiarity" a social barrier to advanced education. The ideology of electronic egalitarianism fails in the face of skill deficiencies, segregating the knowledge building process even more definitively than a lack of books or great physical distance. The results may empower the eventual users in an "informating" fashion, as Zuboff suggests, to explore and create new knowledge through the network links. On the other side, it may drive the marginal users into a rote practice, rule-based approach with limited participation—an "automating" implementation. Finally, the numerous dropouts from the electronic class may be left even further behind in any competition to develop the computer skills that are becoming more widely requisite for work in the contemporary economy.

Zuboff (1988) describes two alternative strategies for technology development and implementation in industrial contexts that are relevant to our educational contexts. For Zuboff, the introduction of new electronic technologies occasions the transformation of learning and its social relations, but technology deployment can follow either of two paths, leading to very different consequences. In the first or "automating" strategy, ignoring the information-creating character of new technologies leads implementers to replicate preexisting patterns of work on computers, essentially mechanizing or automating learning. Examples in pedagogy include many drill and tutorial systems as well as broadcast information like remote video and electronic bulletins. Such systems tend to enforce standardization and promote the isolation of users. As McDonald shows, these systems can also foster individual or group alienation from the learning process. This strategy represents a static approach to knowledge management, where the teacher says to students, "We've got it (now and forever); you go get it."

Zuboff's alternative "informating" strategy is to build knowledge-enhancing systems that use new electronic tools to expand skills and increase the exercise of critical judgment while reducing drudge work. This kind of technology deployment has two concurrent social dimensions: deepening understanding through intellective skills building and distributing information more widely. Intellective skill building includes three increasingly complex abilities that progress from (1) recognizing abstract systems of representation in computers as being related to other real-world processes, (2) understanding the logical or reasoning processes represented by them, and (3) manipulating the symbolic systems to create new information or knowledge. This is congruent with a pedagogic aim of building not just static knowledge but dynamic learning capabilities in students. In this second dimension of distribution, the question of knowledge and access is expanded from the ease of tool use, as in our "glass tools" discussion, to include the availability of hardware and software, access to data or information, and the *political* ability to make informed choices about appropriate technologies and knowledge domains to be learned. In Zuboff's informating strategy, by promoting autonomous judgment rather than rote learning, flexible systems can create continuous cycles of learning and improvement. Autonomous control empowers users to invent new connections and to create new context-sensitive local knowledge resources, and these new knowledge connections feed upon each other in a continuous cycle of expansion in depth and scope.

In the teaching case we describe, this potential cycle of local knowledge expansion may conflict with and be blocked by other institutional powers beyond the classroom. McDonald discusses the influence of school economics and the power (or ignorance) of an administrative system, and Bader and Nyce describe the vexing relationship of American cultural values of individual choice in conflict with a teacher's need to control education. In each case, the arena of discussion has opened far beyond the user-to-tool interface to include ever broader social, political, cultural, and economic environments. In this sense, our articles are an ethnographic argument for expanding the theme of "organizational transparency" from Seely-Brown and Duguid (1990) into an examination of the social transparency or opacity of technologies. Ethnography opens the tool transparency argument to the broader politics of access through systematic examination of the relationship between technologies and their social contexts. Our arguments cut across the multiple dimensions and complex interactions of users' encounters with new technical systems and the transformations of power and social relationships that these encounters engender.

Bridges to the World of Work

By casting these arguments in terms common to new technology implementation in industrial as well as educational environments, I hope to create a conceptual link between our daily teaching practice and the practice of anthropology in other settings. Our issues are relevant to many anthropologists working on the design or implementation of new electronic technologies, and we offer some special insights on that process.

• Issues and controversy around access, skills and expertise (or deskilling), empowerment, and organizational control are common to all institutions that introduce new technologies.

• Because the teachers/researchers who wrote these studies have intimate *access* and at least partial *control* of the immediate environment of most of these cases, they have observed and intervened relatively freely in the tool use process. Their insights from privileged access can highlight problems and detail features that are either more obscure to a consulting practitioner/advisor or are proprietary and cannot be published.

• The volume presents detailed and *critical* ethnographic examples of new technology introductions. These examples can provide other practitioners food for thought and conceptual models for similar environments and help validate and strengthen arguments for knowledge-enhancing systems.

Our cases can thus shed light, for example, on the work of development anthropologists on new technologies for empowerment; they help rethink training and implementation processes in government agencies or the private sector; and some comparisons draw analogies to all electronic tool implementations—not just teaching tools. Our aim is to simultaneously cast data from the educational arena in its wider context of technology development and to bring this rich data to bear on the broader relations among tools, users, power, and knowledge.

Hamill and Marchant begin our volume by asking whether computerized tutorials really help students learn. While describing several specific programs, they propose two dimensions for evaluation: the *venue* for use—in class or outside, and the *range* of usefulness—how wide a variety of learning goals the software can serve. Mason Weiss, Metzger, and McDonald discuss "enhancing group memory" through a hypertext system for easy, flexible access to complex information. This system is coupled with an electronic conferencing tool that allows remote access and group discussions. Edwin Segal tackles the problems of using video telecourse materials when the curriculum is out of the instructor's control.

While these articles look primarily at the classroom, the next two articles open the scope of argument to wider political contexts. McDonald's critical appraisal of distance learning with a computer conferencing tool strongly distinguishes rote or recipe learning from *learning* to learn—a capability that has largely eluded his junior college's technology implementation. Bader and Nyce discuss a hypermedia tool, as well as the clash of teacher's intentions to "guide" with the tool designers' intent to offer "free choice" in how students learn. In addition, two commentators provide critical appraisals of the issues: Anna Hargreaves, an instructional design specialist with a decade's experience in the computer industry, and Greg Truex, a pioneer in electronic media for anthropology.

Finally, we hope that our volume will bridge the work of two major constituencies in NAPA: practitioner/consultants and teacher/academics. By highlighting common issues such as tool design, use, and power in the knowledge building process, we can expand the dialog between teachers as practitioners and other anthropologist-practitioners in our wider professional community.

Notes

Acknowledgments. Our papers grew out of a session at the 1990 AAA meetings co-sponsored by NAPA and CAE: "Closing the Educational Access Gap: Alternate and High-Technology Teaching Methodologies and Their Implications," organized by Jim McDonald and myself. My thanks to all the other authors in this volume for their comments at the session and afterwards, and especially to Jim McDonald and Anna Hargreaves for their close readings. Thanks also to the *NAPA Bulletin* editors and reviewer Jeff Nash for their comments and advice. Our special appreciation is due to reviewer Barbara Russell, who gave a detailed editorial critique of our entire volume as well as its individual papers. My appreciation also extends to the School of American Research in Santa Fe for providing the opportunity as Resident NEH Fellow during 1992–93 to complete work on this manuscript and volume.

1. "What You See Is What You Get." A WYSIWYG system represents the final appearance of a printed document on the screen, giving direct visual feedback. Command language systems cannot do this easily, because of the commands interspersed throughout the text.

2. "Informate" is Zuboff's neologism for her conceptual distinction from "automate," a term that she defines specifically as a Taylorist system of control in new electronic technologies.

References Cited

Barley, Stephen R.
    1986   Technology as an Occasion for Structuring: Evidence from Observations of CT Scanners and the Social Order of Radiology Departments. Administrative Science Quarterly 31(1):78–108.
    1990   The Alignment of Technology and Structure through Roles and Networks. Administrative Science Quarterly 35(1):61–103.
Bjerknes, Gro, Pelle Ehn, and Morten Kyng, eds.
    1987   Computers and Democracy—A Scandinavian Challenge. Brookfield, VT: Avebury.
Ehn, Pelle
    1988   Work-Oriented Design of Computer Artifacts. Stockholm: Arbetslivscentrum.
Norman, Donald A.
    1988   The Psychology of Everyday Things. New York: Basic Books.
Seely-Brown, John, and Paul Duguid
    1990   Design for Implementability. Paper presented at the conference on Technology and the Future of Work, Stanford University, March 28–30.
Zuboff, Shoshana
    1985   Automate/Informate: The Two Faces of Intelligent Technology. Organizational Dynamics (Autumn):4–18.
    1988   In the Age of the Smart Machine. New York: Basic Books.

Articles

Interactive Courseware in Anthropology Classrooms

James F. Hamill and Linda F. Marchant

Introduction

Electronic technology, in the form of personal computers and educational software, presents new classroom opportunities to explore anthropology. However, as with all technological innovations, potential benefits must be weighed against the costs of implementation. In this article, we explore several aspects of how teaching and learning anthropology may be altered with the introduction of these new approaches.

The traditional format used in most college classes emphasizes symbolic learning modes in which students passively listen to lectures or read texts; experimental and exploratory modes of learning where students actually manipulate information are deemphasized. We believe that computers may offer a means to modify this traditional format and foster other learning approaches. The computer's capacity to store and process information may extend classroom learning possibilities. Here we report on the use of personal computers and anthropological software in a number of introductory and advanced undergraduate anthropology classes.

A number of considerations had to be dealt with as we began to use personal computers in some of our lower-division courses. It was by no means clear if the hardware itself could be routinely moved in and out of classrooms. A further issue was availability of both hardware and software outside the classroom for student assignments. Conversely, we were not blindly convinced that such use of computers would improve classes from the teacher's point of view.

On the surface, the issue of whether the classroom use of computers and anthropological software gives students a better grasp of anthropology than traditional methods of instruction remains open. We believe, however, that stated in this way, this issue is a nonquestion. Teachers have a wide variety of technological aids (e.g., chalkboards, overhead projectors, films, and videos) available to them, and computers and educational software are just another set of items in the list. Instructors use these devices when and where they perceive the technology will benefit the class. The real question then is not whether students learn anthropology better with this or any particular classroom technology; rather, it is *how can the technology be put to effective use?*

In our experience, we found that the current models of personal computers and screen projection systems are small enough, inexpensive enough, and "friendly" enough that the hardware itself is not a significant obstacle to using computers to help teach anthropology. Instructional materials (courseware) and ways of incorporating this material into classroom activities are less well developed. We therefore concentrated our efforts in finding appropriate courseware, becoming familiar with its operation, and developing and incorporating it in our classes.

Available Software

We found that two general dimensions defined the utility and quality of the available anthropological courseware. The first dimension, appropriateness of venue, divides software between programs best used in classrooms as lecture aids and those most suited for out-of-class assignments. The second dimension, range of use in anthropological issues and concepts, divides the software between programs that are narrow or limited in scope and wide-ranging programs that can be used as "raw material" for a wide variety of pedagogical goals.

When we first tried using computer courseware for out-of-class assignments, we found that many of the undergraduate students in our classes had little or no experience with computers, and therefore most software had little intrinsic lure. "Assignment" courseware, therefore, should work in the context of an uninformed and perhaps intimidated user. It should be small. Programs that require more than one disk will probably be so complex that most student users will focus only on running the program and miss much of the anthropology it is intended to illustrate. For the same reason it should be "friendly." Programs that require a lot of data entry or complex command syntax will burden students with mindless detail at the expense of important concepts. Finally, good assignment software should be structured so that students interact with the anthropological substance of the assignment, while the computer is in the background enhancing that interaction.

"In-class" software should be judged on the basis of how well it enhances the concepts and information delivered in lecture. This makes complexity (or simplicity) and size relatively less important than it is for assignment courseware. The programs can be as large and complex as necessary, as long as the instructor is familiar with their operation and facile with their use. While program complexity is not so important, clarity of display is. Generally, these programs are used in a classroom, and the outputs are projected onto a large screen. The simpler and clearer those screens are, the better they are. So good "in-class" courseware will feature a series of logically connected screens, each of which is relatively clear and uncluttered and contains a modest amount of information.

The computer's advantage, like interactive video, lies in its ability to store and manipulate information. Film and video are excellent ways to present narrative data, and chalk and slate are much easier to use than even the most friendly computer. The computer, however, can put large amounts of information at the instructor's finger tips and process the information quickly and accurately. Therefore, most good courseware puts data and some processing capacity in the package. The software can be more or less rich with respect to the amount of data it contains. Some programs have only limited data and are intended to exemplify or illustrate only a narrow range of ideas, while other programs provide a lot of data that can be used with respect to a wide range of anthropological concepts.

The dimensions of "appropriate venue" and "range of use" describe attributes of courseware but do not sort the materials into mutually exclusive sets. No instructional software—at least that we have seen—is only appropriate for assignment or any other single use. Similarly, all of the programs that we have found and/or tried to use fall somewhere on a raw material scale, but none is only data to be used with no constraints or only useful for a single instructional purpose.

These two dimensions are continuous scales and not categories, but it serves our purpose here to treat them as if they were categories. When we do, it is possible to look at the courseware we have used and see how each program fits into one of the four boxes shown in Table 1.

Table 1
Anthropological Courseware by Categories

	Venue	
	In Class	Assignment
Narrow	Kinship	Fugawiland
Range of Use		
Wide	MAPSTAT	Cultural Diversity

Fugawiland

Fugawiland (Price and Gebauer 1989) is a good example of "assignment" courseware that is relatively narrow in scope. This archaeology simulation has design features that make it one of the more innovative pieces of courseware available. It assumes that the individual student uses it outside the classroom, and it requires that users interact with the program. It is not a textbook on computer screens; rather it is a problem to be solved. The computer is a tool that allows students to solve the problem in a systematic, and perhaps enjoyable, fashion.

Fugawiland simulates important concepts in archaeological research. The program defines an area in what is now Wisconsin, as well as a "protohistoric" people called the Fugawi. Through "excavation," students find out what the Fugawi ate; where they gathered raw materials for tool making; something about how they organized themselves; and how they used the major resources in their territory.

The program provides a map of the area that shows some major geographical features and the location of 20 or so "known" Fugawi sites. The geographical features include rivers, the Lake Michigan shore, hill country, flatland, and sources of obsidian and copper. The student "excavates" sites and analyzes the findings to determine the site use pattern, subsistence base, settlement pattern, and cultural affiliation of the site makers. When completed, the program tests the student with a series of multiple-choice questions about the "Fugawi."

To succeed, students must make strategic decisions regarding excavation, because only ten of the 20 sites can be used to solve the problem. Early in the simulation, the student is informed that after the tenth site has been excavated, research funds are depleted and no new sites can be opened.

The program provides some analytic tools for students to use. A regional map shows unexcavated and excavated sites. The latter have site maps that display artifacts and features. The program provides descriptive statistical tools such as frequencies and histograms (bar graphs showing frequency distributions of artifacts).

There are some problems with the program. It is not particularly forgiving. If the user makes a mistake, the program may abort and not start up again. The most common mistake involves security of the program. When the program starts, it asks the student user to enter his or her first and last name and student number from the keyboard. On all subsequent start-ups, it asks the student to reproduce this data. We have had problems getting the program to recognize a match, so it has been difficult to use the simulation more than once. Printing on-screen graphics to include with students' final reports on Fugawiland can also be troublesome. Users can only print Fugawiland's graphical displays by entering into DOS graphics mode (i.e., C: graphics [Return]) before booting up the Fugawiland program.

It is also difficult to get around inside the program. Each site that is "excavated" takes the student through a series of text screens that cannot be bypassed. Despite these problems, the program is innovative, easy to use, and structured in a way that requires the student to interact with anthropological concepts. Students cannot be passive; rather, they must make strategic decisions from the beginning of the simulation.

Cultural Diversity

Douglas White's Cultural Diversity (1985) can be used, like Fugawiland, for out-of-class assignments, but because it ranks high on the "raw material" dimension, it is adaptable to a wider variety of uses. Cultural

Diversity is a data set packaged with a program called MAPTAB, which performs various operations on the data. It is available from several sources, including at least two versions in the electronic journal *World Cultures* and another from Wm. C. Brown Publishers. The data are 185 variables coded for the standard sample of 186 societies. The list and codebook are easily extracted from the program itself. The MAPTAB utility is a program that manipulates the data in several ways. It is designed for use with other data sets. Users can generate their own data sets or use precoded information, such as paleontological and anthropometric data (Beals, Smith, and Dodd 1987) and information on magico-religious practitioners (White 1986).

MAPTAB maps variables across the planet and performs statistical operations. You can choose all societies, predefined subsets by five continental areas, or individual societies. Users can select which variables to consider, and some versions of the program provide "prepackaged" bundles of variables that deal with major cultural institutions, such as economic structures, sexual behavior, and social organization. A few keystrokes let you view the codebook for any variable (Table 2 is the codebook for variable 77 taken from Cultural Diversity) and show the attributes of the variable for each selected society on a planetary map. You can also display the marginal frequency (expressed as frequency or percentage) of any variable for the sample, along with a cross-tabulation of any two variables. The cross-tabulations can be displayed as frequencies and/or percentages.

The program is self-contained and simple enough to make assignments in introductory-level classes. The entire Cultural Diversity package fits on one 360-kilobyte floppy disk. This includes the data, the codebooks, and the MAPTAB utility. This is an important advantage over the MAPSTAT package, which takes five disks, if the package is used for assignments. The program is menu driven and user friendly, although the screen displays are somewhat cluttered. It shows the variable codes and performs statistical operations from inside the program.

The maps, at best, are crude. Because students are more familiar with maps than cross-tabulations, they tend to emphasize the mapping features when they use the program and sometimes get frustrated with them. Apart

Table 2
Cultural Diversity Codebook for Variable 77

---

77. Role of the Older Generation in Arranging Marriages [Var 603 Study 15] (1st Marriages Only). Missing Data

- Missing Data
1 Males monopolize arrangement
2 Both males and females participate, males have more say
3 Both participate, and with roughly equal say
4 Both males and females participate, females have more say

---

from the quality of the maps, the program is flexible and powerful. It is an excellent resource wherein students interact directly with high-quality anthropological data. This package allows students to set up problems that demonstrate the range of cultural variability and the interrelatedness of various aspects of culture.

The package is relatively inexpensive. Single copies are available from Wm. C. Brown for about $35, and site licenses for the package are available from all of the vendors and range in price from $200 to $400.

MAPSTAT

MAPSTAT is another *World Cultures* package that gets high "raw material" points. It is so large and cumbersome, however, when compared to the Cultural Diversity package that its use should probably be restricted to the classroom. It is, however, a powerful tool for demonstrating cultural relationships, and it has significant pedagogical use in the classroom.

MAPSTAT is actually three programs that interact with one another on a large and growing body of cross-cultural data. When published in 1988, it had over one thousand variables coded for the 186 societies in the standard cross-cultural sample, with the codebook for each variable. The programs, data, and codebooks take up five double-sided, double-density 5 1/4-inch diskettes. Additional variables with codebooks continue to be added. The three programs in the package are MAP, SORT, and STAT.

- The MAP utility displays cultural variables on six high-quality continental maps.
- The SORT utility sorts and recodes any of the more than 1,000 variables in the package, according to user needs.
- The STAT utility produces various statistics (e.g., chi-square, Pearson's *r*) showing relationships between variables that the user selects.

We use the package in various ways as a lecture aid in conjunction with a screen projector. One application is to show the frequency and distribution of different cultural traits across the planet. The maps in the mapping program are clear and project well. The mapping routine will handle two variables at once and will perform and display correlation statistics on the chosen variables. The package can also be used in class to answer questions about interrelationships between aspects of culture (e.g., Is the level of cultural complexity related to the pattern of postnuptial residence?).

The power of the package makes it large and, therefore, somewhat awkward. It is difficult to sift through over a thousand variables in over 20 codebook files in the context of a single lecture. These small research projects require familiarity with the programs and work best if the instructor prepares before class. The instructor needs to use the program's various utilities to sort out whatever information he or she may want to use on a day-to-day basis. The program is friendly, and it is relatively easy to tailor the information to course and lecture needs.

Kinship

Kinship (Ottenheimer 1988) is an example of narrow application courseware that is designed to be used in class with a screen projector. It consists of a series of seven modules that use graphics to demonstrate various aspects of kinship and social organization. The seven modules cover important concepts and axioms in marriage and kinship systems. For example, the introductory module covers kinship symbols and how they are used; kinds of cousins; marriage systems; fundamentals of unilineal and bilateral organization; and rules of postnuptial residence. An eighth module sets up printers for graphics printouts.

The screens are uncluttered, and an especially attractive feature of this program is the use of animation to demonstrate many of the processes at work in kinship systems. Like Fugawiland, the program is rather narrow, with a limited set of uses. It also shares similar technical limitations. Printouts of the graphics are problematic, and, as with Fugawiland, it is hard to get around inside the program. For example the introductory module has over thirty screens that cover a wide variety of topics. If you wish to start on a topic other than the first screen, you must cycle through all the previous topic screens. This is needlessly cumbersome; one solution would be the addition of a second level of hierarchy that breaks up each module into subtopics. The program was published in *World Cultures* and is also available through Wm. C. Brown (Ottenheimer 1989).

These four programs represent the sort of courseware that is available today. Two of the programs are more appropriate for in-class use (MAPSTAT and Kinship), and the other two programs can easily be used for out-of-class assignments. Two are narrow application packages (Fugawiland and Kinship), while MAPSTAT and Cultural Diversity provide the raw material to teach a wide range of anthropological concepts. We have used three of the four in various class contexts.

Classroom Computer Use

In Spring 1990, we did a pilot of classroom computer use. Neither of us were teaching a class of appropriate content or level to begin our research. We asked a colleague to let us use her class for this first effort. The course is titled "Peoples of the World" and is a lower-division introductory world ethnography class. We suggested a Cultural Diversity assignment to the instructor, and she agreed. The class itself was not organized around computer use, and the assignment we chose was not part of the class structure, nor was it directly related to class material.

The assignment was to use the Cultural Diversity program to see if some relationship, that the student defined, existed between aspects of culture. To complete the assignment, students wrote a brief report using an outline we provided. The assignment was optional, and the instructor defined its value.

We made and graded the assignment, provided the necessary instructions on using the program, and set up the program in the university computer lab for student access. We used four class periods for the pilot project. The first two were devoted to showing students how to use the Cultural Diversity program. We brought a computer and overhead projector pad into class and went step-by-step through the program features. We provided a handout to show each keystroke sequence and its effect. Students used this to follow on paper what we were doing with the machine and as a user's manual when they were on their own. We handed out the assignment at the conclusion of the second session. Students were given two weeks to complete the assignment. The third classroom session followed up on our computer instruction a week after the assignment was made. In the fourth session, we asked the class to evaluate the process and assignment.

During the Fall semester of 1990, we took the next step and organized two classes around computer courseware. One course was the introductory-level world ethnography course that we used the previous spring and the other course was a junior/senior-level social anthropology course. In both courses we used Kinship and the MAPSTAT package as lecture aids and the Cultural Diversity package for research assignments.

Because we have only one computer set up for course use, we decided that only one of us, Hamill, should teach computer-based courses. We also felt the computer should be as familiar or "transparent" as possible, so it was used in both courses on a daily basis.

This daily use was structured around hypertext outlines of the course material within which we embedded the various courseware packages. Class preparation involved using an outlining program for the day's lecture. The outline included the material to be covered and whatever program calls would be necessary for that day. In other words, if the lecture required courseware in addition to the outline, the commands to execute those programs were included in the outline.

In class we used the Kinship program, and the MAPSTAT and Cultural Diversity packages. Both classes also used the Cultural Diversity package to complete research assignments. The assignments were similar to the pilot project of Spring 1990. Students used the data set to determine if some relationship existed between different aspects of culture. The assignments differed from the first ones in that they were more narrowly defined; they were required; and they were worth a substantial portion of the final grade. For the upper-division course, the assignments were also structured to aid in the research for the required term paper. We devoted a week of class time to software instruction, as well as several hours in the university's computer lab. One class meeting had a guest lecture on the use of the statistics in the package. Additional time was invested in instructing the computer lab assistants in how the package works.

Conclusions

We are still clearly in the process of exploring computer use in anthropology classrooms. However, we have already learned some valuable lessons—some logistic and some pedagogical. We have demonstrated, at least to our satisfaction, that this undertaking is possible. But there are several caveats. Two important issues that must initially be considered are one, faculty preparation and, two, support systems.

It is crucial for faculty to be well acquainted with both required hardware and software. We need to be at ease with the technology and applications. If we are to succeed in using these less familiar materials, then students ought not to be distracted by technical problems. Rather, their interest should be held by the content of the course.

This suggests an additional burden on the faculty member. We all prepare our class content. With these new teaching approaches, class preparation is somewhat different. We will need to tailor data bases and applications to our existing lecture framework and scheduling constraints. We will also need to carefully balance the time we devote to preparing anthropological content versus technological requirements.

Support systems can and will make or break these endeavors. Obtaining, maintaining, and retaining the actual hardware is no small feat. Computers and screen projection systems must be committed virtually full-time to these efforts. Thus far we have been compelled to move our equipment in and out of the classroom on a daily basis—not the ideal situation. Not only does it create logistics problems—elevators, carts, and so on—it results in significant wear and tear on the equipment. This, in turn, will result in additional maintenance.

Preliminary information on student perception of using computers to help teach anthropology shows some variability but gives reason to continue. In our first try, where we were guests in another teacher's course, we made an open-ended assignment. Only three students from a class of nearly 80 even attempted the assignment. Student evaluations showed that the benefit from doing the work was too small to justify the effort (the instructor gave only five points out of 200 to the task). Of the three students who attempted to complete the assignment, one reported that he had learned something new about cultures, another reported that he was not able to discover anything he did not already know, and the third said that she could not make the program work.

These results told us that for anthropology courseware to be an effective teaching tool, it had to be relevant to course material, and computer use had to be integrated into classroom activities. On our next try, the computer was in the classroom every day, and students were required to complete assignments using the Cultural Diversity data set. Evaluations of this teaching structure indicate that, in general, students did not feel that the daily class use of the computer interfered with the anthropological content. One of the questions we asked when we evaluated the computer project was whether the in-class use of computers enhanced or detracted from the course. We gave the evaluation to both of the classes we organized around daily computer use, and a total of 54 students replied. Fifteen (27%) replied that the computer detracted from the course and 19 (35%) replied that it enhanced the course. The remaining 20 were neutral on the subject.

On another evaluation that did not address computer use, about 40% of the students (21 of 49) volunteered comments on the in-class use of computers. Of those 21, 16 were negative (e.g., "The course seemed to drag, especially when using the overhead computer or talking about the Cultural Diversity data base," "the course is fine except for the computer nonsense," and "the Cultural Diversity Database Program sucks") and five were positive (e.g., "Use of computers in class is a unique and organized method of teaching," "the computer outlines were helpful," "the computer assignments were effective for . . . learning of certain variations in societies throughout the world").

We conclude from these responses that the students were aware of the computer as a computer. The machine was not transparent, and our strategy of having it in the classroom almost every day worked against making it so. Instead of a "glass tool" that helped students learn, it became a focus of attention that sometimes pulled students away from the anthropology. If transparency is not achieved through exposure, it may come about through the logic of course material. Instead of "inoculating" students to the computer by showing it to them every day, introduce them to the advantages of computer-based information by using the machine only in classes where it is relevant to the material.

If these projects are to continue, they require institutional support. Departments need to commit faculty time to develop these methods of instruction; funding for hardware and software is required; and, ultimately, we will need to demonstrate that these new methods are successful. We will have to show how students benefit from these approaches. We need to develop better ways to use the computer so students learn more, have a greater depth of understanding, and are able to explore anthropological materials in ways that are difficult or impossible without it.

References Cited

Beals, Kenneth, Courtland Smith, and Stephen Dodd
    1987   Paleontology and Anthropometry. World Cultures 3(3).
Heine, Gerald
    1988   MAPSTAT. World Cultures 4(2).
Ottenheimer, Martin
    1988   Kinship. World Cultures 4(1).
    1989   Modeling Systems of Kinship and Marriage. Dubuque, IA: Wm. C. Brown.

Price, Douglas, and Anne Gebauer
    1989   Fugawiland (version 2). Bakersfield, CA: Mayfield.
White, Douglas
    1985   Cultural Diversity. World Cultures 1(3).
    1986   Magico Religious Practitioner Data. World Cultures 2(3).
    1987   Cultural Diversity. Dubuque, IA: Wm. C. Brown.
    1989   MAPTAB (rev. 1). World Cultures 5(2).

Hypertext Indexing Applied to Computer-Mediated Conferencing and Teaching: An Aid to Group Memory

Audrey E. Mason Weiss, Duane G. Metzger, and James H. McDonald

Computer conferencing provides opportunities to explore the relevance and utility of remote, asynchronous[1] instruction and learning. Individuals participate in a computer conference by dialing into a mainframe computer with their own personal computer and modem. Once connected to the mainframe, they can join conference(s), read previous entries, add their responses, create new entries, and turn in completed projects. A computer conference on a given topic unfolds over time as participants exchange ideas and data and request more information and clarifications from one another.

If a conference has many active members, three problems arise. Within a reasonably short time, there may be hundreds of messages that layer up chronologically. A participant is faced with the task of remembering who said what and when they said it. Furthermore, participants must also attempt to remember how the discourse of a particular conference developed and transformed over time. Finally, new participants are faced with too much information to adequately review or understand past discussions. This article examines these problems in detail, using our experience with the BESTNet computer conferencing system as an example. We will then offer a solution to these problems that blends basic ethnographic methods informed by ethnosemantic theory with hypertext software. The result is a custom information management system that functions as an *aide-mémoire* to help computer conference participants come to grips with the problem of information overload: too much information to remember and too many people to keep straight over time.

BESTNet and the Problem of Information Management

The computer technology necessary for connecting geographically dispersed sites in a computer-mediated conference is available worldwide.[2] In fact, throughout most of the world, universities are interconnected through their computer mainframes, creating a vast electronic network in the manner of a world telephone system (Arias and Bellman 1987; Bellman 1988; Feenberg1988; Hiltz and Turoff 1993[1978]; Houser and Wallace 1989; Kerr and Hiltz 1982).

An example of a multiuniversity computer conferencing system is the Bilingual English/Spanish Telecommunications Network (BESTNet), which is made up of a consortium of universities in Argentina, Canada, Mexico, and the United States. The institutions share computer conferences, computer-based courses, electronic mail, and data bases that are available to interested students and faculty.

Computer conferencing holds tremendous potential for instruction. It also facilitates the exchange of ideas and data between individuals who are geographically dispersed or have schedules that do not permit them to be in the same place at the same time. BESTNet offers a wide range of conference topics including agriculture, anthropology, botany, engineering, intercultural communications, medicine, psychology, and telecommunications. BESTNet has the advantage, then, of bringing together a diversity of people and experience that otherwise would not have the opportunity to interact.

In our experience, however, individual recall of past dialogue in a computer conference is problematic. Individual and group memory of the conference is faulty. A computer-based course or conference on BESTNet often involves as many as 100 individuals distributed across eight institutions over durations as long as one year. This results in a chronologically ordered sequence of hundreds of entries by numerous participants over the life of the conference. The problem for a participant, then, becomes one of organizing this mass of information in a way that is both useful and meaningful. This problem is further compounded by the chronological structure of the conference itself. Responses to an original entry may not necessarily fall immediately after the entry on which they were commented. Additionally, long strings of commentary that began as a discussion of one theme or topic often *evolve* into another. Simply keeping track of a discussion and its various mutations can pose a major problem for participants. It is helpful to think of these conferences as analogous to an extended, and sometimes unwieldy, conversation.

To better understand the problems facing conference participants, it is useful to examine in more detail the actual structure of a computer conferencing system. In our case, BESTNet uses the widely available Vaxnotes (Digital Equipment Corporation 1986) software, which employs a fairly common conference structure. At any given time, there are a number of concurrent conferences running on BESTNet (e.g., pertaining to anthropology, agriculture, or psychology).

Within a conference, the basic communication structure can be termed "topic and response." A "topic" is an entry that requests or provides new information that is not strongly related to previous exchanges in the conference. A "response" is attached to a previous "topic." For example, a participant in a BESTNet conference on "Ethnobotany" requested information on ethnobotanical remedies for diabetes. Shortly thereafter, another participant "responded" by describing an indigenous plant remedy she collected while doing fieldwork, and another "responded" with a short list of relevant literature. Each entry receives a unique number within the conference, and entries are listed chronologically. Referring to Table 1, the three entries described above are topic/response numbers 5, 9, and 16, in an abbreviated listing of entries from this conference. This exchange sparked a lengthy discussion of traditional remedies that continued over several months, involving faculty and students from many different universities in the BESTNet consortium. A conference, thus, grows through the linear addition of sequential topics and responses. Earlier entries, as they become remote in time and context, may lose their value at later stages of the conference discussion.[3] Such a linearly arranged sequence of entries quickly passes beyond both the individual and collective memory of the participants as their recall falters.

The designers of computer conference software have recognized these problems and have taken steps to help conference participants manage these huge information data bases. Vaxnotes incorporates a number of memory-aid devices to mitigate the information decay described above. These include directories that sequentially list entries and their titles, keyword indexes, and the ability to conduct searches within the conference by numerical identification (each entry has its own unique number), date,

Table 1
Title Examples

Directory Conference Topics: Ethnobotany [→ Titles ←]
Created: 6-MAR-1990 10:36 18 topics Updated: 16-OCT-1990 15:46

Topic	Author	Date	Repl	Titles
3	UCSVAX::BESTR1081	22-MAR-1990	0	Baja CA Diabetes Plants
4	UCSVAX::BESTR1081	22-MAR-1990	0	Baja/Mexico Medicinal Plants
5	UCSVAX::BESTR1081	22-MAR-1990	0	I Need Diabetes Plants!
6	UCSVAX::BESTR1081	22-MAR-1990	2	Request for Trad. Lore
7	UCSVAX::BESTR1099	22-MAR-1990	1	Online Ethnobotany Biblio.
8	UCSVAX::BESTR1081	22-MAR-1990	0	Ethno Biotechnology Biblio.
9	UCSVAX::BESTR1090	23-APR-1990	3	Plant Remedies for Diabetes
10	UCSVAX::BESTR1099	8-MAY-1990	0	Tzeltal-Tzotzil Ethnobotany
11	UCSVAX::MCCURDY	29-MAY-1990	0	Fenway Park
12	UCSVAX::BESTR1081	16-JUN-1990	0	Summer Absence
13	UCSVAX::BESTR1081	27-JUL-1990	1	Ethnobotany Curriculum Comments
14	UCSVAX::BESTR1081	27-JUL-1990	0	See Note 13.0
15	UCSVAX::BESTR1081	18-AUG-1990	0	Redbud Use-Anderson
16	UCSVAX::BESTR1081	18-SEP-1990	1	Data on Diabetes Plants

author, title, and key words or phrases within the entries.[4] A participant can simply read this information, as well as save it, by filing it electronically or printing a hard copy.

One of the most commonly used methods for recalling the content of entries is by title. As a consequence, part of the task of building a viable re-call structure for a conference rests with the individual BESTNet member. He or she is asked to provide a short, descriptive title that ideally represents a phrase-long abstract for the entry. Good titles are critical because they are among the most frequently used memory aids.

Unfortunately, many BESTNet members are not experienced in creat-ing useful titles for their entries. From our experience, titling canons are not widely understood or shared among members, nor is the function of im-proved recall necessarily a high priority among members when they title their entries. Table 1 illustrates a number of recall devices used by Vaxnotes that appear in a basic conference directory. Directories list entries sequen-tially, providing information on each entry's number, date, author, number of replies to the entry, and title.[5] A striking feature of this directory is its lack of continuity and clarity (and thus its use as a recall device) of entry titles.

Applying Ethnosemantic Theory/Methods in the Management of Computer Conferences

The question that we can now pose is: how can a computer conference be made more accessible and useful to its participants?[6] As a point of entry, it is helpful to view a conference as the result of the exchanges of informa-tion between members of a group. The smooth flow of communication is grounded in the development of consensus concerning the purpose and direction of the group activity, as well as the building up of a common pool of knowledge among participants. Consensus produces common expec-tations among participants as well (Romney and Weller 1984, 1989; Weller and Romney 1988). These expectations can be treated as if they were a be-lief system that can be studied and understood using ethnographic meth-ods. Because a conference is a relatively tightly bounded system, its mul-tiple entries represent a more or less coherent body of belief—a "cognitive residue" of its members (Truex and White 1988).

Two important implications follow from the treatment of a computer conference as a belief system. First, a group culture is constructed through the conference process. Second, it should be possible to write an ethnog-raphy of the conference/group culture to sharpen members' understanding of the belief system they have created and continue to help fashion.

We have found HyperRez[7] (Larson 1990) hypertext software to be a useful tool in accomplishing this task because it accomplishes the ethnog-rapher's job of accessing and representing the domains of knowledge em-bodied in a conference, how each domain is internally constructed, and how they relate to other domains of knowledge within the conference.

Taken as a whole, HyperRez, as an analytical tool, reveals the structure of knowledge for the conference. Hypertext analysis results in an operational model of a conference that is similar to those developed in ethnosemantic analysis associated with cognitive anthropology (Black and Metzger 1969; Frake 1969; Kay and Metzger 1973; Metzger and Williams 1963; Tyler 1969; von Glascoe and Metzger 1979). "It is a way of discerning how people construe their world of experience by how they talk about it" (Frake 1969:29). From this ethnosemantic perspective, cultural knowledge is arranged in a number of domains that are composed of hierarchically related categories. As we move down the hierarchy, categories become increasingly exclusive. The overall form is a tree-like structure of cultural knowledge that resembles a network of hierarchical interconnections and relationships.

Hypertext provides a tool for generating a structural description of the culture/text represented in a computer conference. It can theoretically link any chunk of data in any format (e.g., text, graph) with any other chunk. But more importantly, a metastructure of the hypertext arranges information into hierarchically organized categories. This metastructure is generally designed, at least initially, by an expert.[8] He or she will design the hierarchy of categories in the system and the "jumps" or pathways from one level to the next.[9] Student users then add their content to these categories and may also critique them as well. As a group, users are encouraged to redesign those categories when appropriate, based on their accumulation of knowledge and experience. Ideally, student groups will ultimately reach the point where they could jointly create their own metastructure (i.e., classification systems) and construct their own meaning(s). Thus, what is created on a hypertext system is not a static body of knowledge, but one that is organic, constantly growing and changing over time.

User creativity is also encouraged by the multiple choices available to a hypertext user in terms of files viewed at each categorical level of the system, as well as the pathways followed through the hypertext program. Specifically, each category of the hypertext metastructure will contain multiple files and the hypertext pathways ("jumps") that provide users with the tools to manage information—to probe and question it—in ways that are not possible on a regular computer conferencing system.

In sum, hypertext browsing functions analogously to the analytical principles used in ethnosemantic analysis, with the added benefit of presenting users with choices that lead directly to information without prior knowledge of its existence (Larson 1990). The hypertext viewing of electronic, as opposed to printed text, offers the reader new ways of putting textual information together that are emergent and creative in terms of their composition (Crane 1991).[10] Therefore, hypertext is not just a better way of managing large amounts of information. It can be used as a discovery procedure to find new and unexpected information or new and unexpected connections within a pool of information, a function not generally accomplished by traditional textual data base searches that employ Boolean methods (e.g., similar to what you would use in an on-line library search) (Larson 1990; Metzger 1988, 1989, 1990a,b,c; Metzger and Beltran 1990; Metzger and Weiss 1989; Romney and Weller 1989; Weiss and Metzger 1990). Finally, HyperRez is easy to learn and use so that its utility is not restricted to the computer-sophisticated user. As a consequence, the ethnosemantic analysis of the structure of knowledge of a computer conference could be carried out by a single individual for the rest of the group; or, given the user-friendliness of the software, each participant could generate an analysis of the conference.

Examples of Hypertext Management of a Computer Conference

How, then, does hypertext work? It is helpful to begin with a simple example that illustrates the utility of HyperRez. Tables 2 and 3 represent examples of the most inclusive level of categories in a computer conference on "Bicultural Diversity." Both examples are identical to what would be displayed on the HyperRez user's computer monitor. The computer's up- or down-arrow keys allow the user to move the cursor up and down through the bracketed file names (e.g., <filename>), highlighting them with each move. The right-arrow key will open (i.e., "execute a jump" or move between levels in the hierarchy of categories in the hypertext metastructure) any highlighted file of the user's choice. The left-arrow key will return the user to previously viewed screens in the reverse order of their viewing.

Referring to Table 2, START.TEXT provides the logical starting point for analyzing the content of the conference, and it also provides a brief description of the conference as well. By moving the cursor onto the bracketed <index> that appears in the first paragraph of START.TEXT and pressing the right-arrow key, a user can view the index to the conference on "Bicultural Personality." The index, represented in Table 3, lists all of the categories of information in the conference. Again, running the cursor up

Table 2
Partial Start Text for the Bicultural Personality Conference

A partial Start Text for the Bicultural Conference. Names included in brackets < >, are file names to which a browser can "jump."
START.TEXT <start.txt>
The Heuristic Domains of Thought for First Organization of the Texts
<domain1> <domain2>
This first partial ordering of the texts is categorized into 27 nonmutually exclusive domains. <index> This partial ordering has the character of a developing classification. As new theme domains appeared during the work they were added to the classification. Retrospective reorganization will be necessary. <domain1>

Table 3
Index for the Bicultural Personality Conference

INDEX

\<Index\>

  1> Agents/Mechanisms for Social and Cultural Change \<cngagent\>
  2> American Indians in Bicultural/Border Perspective/Context \<amerind\>
  3> Assumptions \<assump\>
  4> Bicultural Cognition \<bicog\>
  5> Bicultural Conflict/Violated Expectations \<biconfli\>
  6> Bicultural Consciousness/Perception \<bicon\>
  7> Bicultural Persona \<biculper\>
  8> Bicultural Practise and Experience \<biculexp\>
  9> Bicultural/Border Books/Periodicals/Publications \<bipubs\>
10> Bilingual/language/linguas francas \<bilang\>
11> Biographies of Persons Related To the Texts \<bio\>
12> Border Consciousness \<borcon\>
13> Border History \<borhis\>
14> Border/Bicultural Places \<urban\>
15> Conference Working Norms/Practices/Politics \<cnfrnorg\>
16> Culturally Distinct Groups Recognized \<culgrps\>

---

and down these categories and pressing the right-arrow key on the desired category will open it for further exploration.

If a user had just joined the conference on "Bicultural Personality" and wanted to know the names of other members, HyperRez will generate a list. Scanning the list of categories in Table 3 reveals that number 11, "Biographies of Persons Related to the Texts \<bio\>," should provide information about fellow participants. Press your down-arrow key and highlight \<bio\>, and then press the right-arrow key. The screen shown in Table 4 will appear. Skip with the down-arrow key through the names, and the execution of the right-arrow key at any of the highlighted names provides you with the texts written by or about that individual participant.

Another more complex example comes from a conference entitled, "Ethnography" (Metzger et al. 1990; Metzger and Ortiz 1989). The conference is associated with a university course on that subject. One of the many categories of information that emerged in the conference concerned "pets." Table 5 shows the considerable number of subcategories that evolve out of the more inclusive category of "pets." Highlighting and opening any of these will reveal the various texts produced by students pertaining to that subcategory.

A final example, represented in Table 6, is a fully expanded START.TEXT from a conference entitled, "CETYS: PSICOLOGIA" (Weiss and Metzger 1990). As noted above, a conference member could browse through the bracketed file names in the right-hand column using the up- or

Table 4
Example of a Second "Jump" to Biography <bio>

```
File: Biographies <bio>
Steve Arvisu <biosa>
Tomas Atencio <biota> <edu>
Armando Arias <bioaa> <bioaal> <biocogaa> <bicogaal> <biclpraa>
<dance> <baptize> <multiper>
Oswald Baca <biobb> <edu>
Beryl Bellman <biobb> <beating> <probbb> <belgibbs>
Victor Hugo Celaya <biovhc> <visit>
Rodolfo Chavez Chavez <biorcc> <visit> <edu>
Alfredo Cuellar <bioac>
Robert A. Devillar <biorad>
Blanca Esponde <biobe> <visit>
Manuel Esteban <biome> <visit> <edu>
Durbin Feeling <borcon5>
Christine A. von Glascoe <biocvg> <hlthsurv>
Felipe Gonzales <edu>
```

down-arrow keys, select a topic of interest, then "jump" to that topical category by pressing the right-arrow key. Once the topical screen appears, the user can "jump" to any individual entry by again highlighting it and pressing the right-arrow key.

Conclusion

Computer conferencing is playing an increasingly important role in education, and in industry as well. When geographic or temporal barriers keep people from being in the same place at the same time, asynchronous technology such as computer conferencing provides a solution to communication problems. Participants can access their conference when *their* schedules permit—it is a flexible solution to the exigencies of time and space (Zimmer 1988).

Computer conferencing also creates a number of problems. Conferences with numerous active participants can end up with far more information than an individual can or would care to remember. Yet, some of that lost or forgotten information might prove critical to current understanding of a conference discussion or to the creation of future topics. Similarly, massive amounts of information make it difficult for new participants to adequately review past discourse. In sum, on BESTNet conferences there is too much information and too many people to keep straight over time. The designers of computer conferencing software are aware of these problems and have designed a number of memory-aid devices into their software. We have argued that these aids are inadequate. Our solution is the custom construction of an information management system that organizes masses of information in an understandable and retrievable form.[11]

Table 5
Pet Themes in the First Ethnography Conference

START/TEXT <start.txt>
1 Classification of Dogs <pet1> <pet2> <pet10> <pet15> <pet20>
   <pet22> <pet26> <pet45> <pet50> <pet60> <pet61> <pet65> <pet66>
2 Pet Names <pet1> <pet3> <pet4> <pet12> <pet29> <pet33> <pet34>
   <pet39> <pet42> <pet46> <pet64>
3 Cross Species Naming <pet4> <pet6> <pet20> <pet23> <pet28>
   <pet31> <pet38> <pet64>
4 Human-Pet Interaction, Petting: <pet7> <pet8> <pet9> <pet10> <pet12>
   <pet14> <pet15> <pet44> <pet62> <pet65>; playing <pet16>; licking
   <pet16> <pet30> <pet38> <pet42>
5 Domestic/Household Animals; xico <pet11> <pet13> <pet29> <pet51>
   <pet62>
6 Sayings, Stories from Dog Observation: Africa <pet16> <pet37>
7 Cultural Definition, Concept of Pet <pet17> <pet18> Dog <pet37>
   <pet40> <pet60>
8 Treatment of Animals <pet19> <pet26> <pet27> <pet28> <pet32>
   <pet35> <pet41> <pet66>
9 Pets in China <pet20> <pet21> <pet22> <pet23> <pet24> <pet27>
   <pet28> <pet32> <pet35> <pet36>
10 Social Theory of Domestic Animals: pets <pet20> <pet21> <pet24>
   <pet26> <pet31> <pet38> <pet39> <pet40> <pet41> <pet43>
   <pet48> <pet49> <pet50> <pet66> <pet67>
11 Functions of Pets: Dogs: <pet22> <pet36> <pet38> <pet40> <pet41>
   <pet48>
12 Cats <pet39> <pet42> <pet43> <pet63> <pet64> <pet68>
13 Spiritual Power of Dogs <pet25> <pet34>
14 Animal Healing <pet42> <pet48>
15 Mourning for Pets <pet20> <pet43>

We suggest HyperRez hypertext software as a solution to the management of information that often clogs active computer conferences. This software allows a participant to quickly identify fellow participants and important categories and subcategories of information that have been discussed in the conference.

We have further argued that the process of computer conferencing can be thought of as the emergence of a group culture or a belief system. Consequently, we should be able to apply ethnographic techniques to help organize, manage, and understand a computer conference. HyperRez software analyzes the information in a computer conference in a manner analogous to ethnosemantic analysis—it reveals the structure of knowledge of a computer conference as a network of hierarchically related categories of information.

The cost of creating this personal information management system is quite small (assuming that mainframe conferencing software has already been purchased). Vaxnotes requires a few hours to learn the basics, and

Table 6
Start.Text of the CETYS Conference (partial)

---

file name <start.txt>

START.TXT: HYPERTEXT OF THE CONFERENCE "CETYS:PSICOLOGIA"
CETYS: PSICOLOGIA

Created: 6-SEP-1990   09:33   16 topics

INTRODUCTION

This is the CETYS:PSICOLOGIA conference in Spanish/English.
This HYPERTEXT version represents one of several pioneering
efforts in the direction of computer mediated
ethnographic/psychological/intercultural/technological
instruction. The entries are summarized in an index.   </aaa/index>
The rationale and agenda of this conference are introduced
in the introduction.   </aaa/1intro>
The story begins, as does all investigation, with people.
PERSONS AND GROUPS:
   Biographies give insights into the people who   </aaa/biografy>/
actively participate in the conference as well as their
experiences with others within the intercultural adventure.
It is hoped that this may represent a kind of "growth chart"
as each embraces the risk to reveal her/his vulnerability and
become transparent to the others.
ORGANIZATIONAL STRUCTURE:
   The Topic and Reply Directory is published for reference.   </aaa/figtprp.f>
   Educational Institutions identifies those formal   </aaa/educinst>
institutions cooperating in this conference.
   Chronological order retains the history of the   </aaa/notes>
conference as it developed.
CONFERENCE PARTICIPATION:
   Getting started is often the hardest part of any new
</aaa/getstart>
adventure, and conferencing is no exception. Several
students share their sense of risk.
   Special interests allow participants to easily identify   </aaa/specint>
and interact with others in the conference who share interests
in common.
CULTURAL INTERACTION:
   Linguistics and language are integral components   </aaa/linguist>
of this conference, since it brings together participants
from different countries and cultural orientations.
Overcoming language barriers is one of the essential goals
of this program.
   Student exchange provides for face-to-face interaction   </aaa/xchnge>
between participants in this conference and the
participating institutions.
   Formal/informal studies are ongoing to provide   </aaa/studies>
students and others with a structured source of
intercultural information.

HyperRez requires only about an hour to learn. As a free shareware package, it is a very low-cost solution to the information management problems faced by computer conference participants. While we have emphasized the educational applications of this hypertext solution, it has considerable potential in industry as well. A useful application of hypertext computer conferences has been as a training and support tool in the area of quality control. Hypertext programmers conduct a "brain drain" on experts, getting them to identify the two or three main customer complaints and product failures based upon a pool of, perhaps, hundreds of individual problems. Other hypertext categories suggested by our experts would center on other related issues: causes of the problem, solutions, how to recognize the problem, related information, and so on. Conference participants would not have to wade through massive amounts of disconnected information concerning quality control problems. Rather, hypertext can be used to manage those discussions from the outset so that problems can be quickly defined, solutions sought, and interventions made; thus cutting training time and facilitating better product support.[12]

Notes

Acknowledgments. We wish to acknowledge support from the Digital Equipment Corporation under grants made to Beryl Bellman and John Witherspoon. We would also like to thank Barbara Russell and Frank Dubinskas for their suggestions and insights on earlier versions of this article.
1. A computer conference is asynchronous when individuals can access the conference at any time and read previous messages and leave their own for others to read at another time. In other words, for individuals to communicate, they do not need to be working on the computer conferencing system at the same time.
2. The community college described by McDonald (this volume) was even able to successfully connect, albeit with some difficulty, with students at several institutions in the states of Russia and Georgia in the former Soviet Union during the Fall of 1992.
3. It should be noted that Vaxnotes does not have an archiving feature so that messages entered before a certain date will automatically be relegated to an archive, making the amount of current information less daunting to the newcomer. A conference participant can, however, issue a command to supress the indexing of entries prior to a given date chosen by the conferee.
4. The ability to create different indexical views of a conference and textual data base searches allow a user to retroactively reorganize a conference up to a point. Users, however, are more likely to find what they already know, but less often discover what they do not know (Larson 1989).
5. In addition to a simple chronological listing of entries, a participant can also call up customized directories by author, date, entry number, or key word to better fit individual needs.
6. It is important to note that there are newer computer conferencing systems, such as Lotus Notes, that come with hyptertext already bundled with the software program. Therefore, our solution is directed to groups using older conferencing software, such as the Vaxnotes example.
7. Information regarding HyperRez software can be obtained by contacting Neil Larson, MaxThink, Inc., 2425 B Channing #592, Berkeley, CA 94704. HyperRez is a memory resident software package. A nonmemory resident version, Hyplus, is also available (Larson 1989).
8. This hypertext software does *not* have an intelligent interface engine that arranges information in a hierarchy of categories (i.e., it is not self-organizing). This must be done through a single expert or group-design process.

9. A simple hypertext metastructure for a computer conference on "Ethnography" might have the following descending hierarchical metastructure in terms of the scope and inclusiveness of categories:

What is Ethnography?

↓

Formal Definition of Field

↓

Examples

↓

Limitations/Anomalies/Problems

↓

Related Topics

10. Others such as Poster (1990) and Finnegan (1991) are critical of the interpretation of hypertext as a technology that results in the emergent, nonlinear production of text/culture. Poster (1990:100–101) contends that hypertext actually reinforces the linearization of information, a stance which counters most other interpretations that emphasize the utility of hypertext (cf. Crane 1991; Landow 1992; see also Bader and Nyce, this volume, and Nyce and Kahn 1992, for an excellent review of the origins and current debates concerning hypertext). What these conflicting approaches underscore is that hypertext or any technology should always be engaged critically as things inscribed with cultural meaning.

11. By providing individuals with a clearer representation of the structure of a conference, their personal recall is enhanced as well as their understanding of the conference as a constructed, group process.

12. Personal communications in October 1992 with John Phillips, an engineer with Digital Equipment Corporation.

References Cited

Arias, Armando, and Beryl Bellman
   1987   BESTNET: International Cooperation Through Interactive Spanish/English Transition Telecourses. Technology and Learning 1(3).
Bellman, Beryl L.
   1988   A Model for User Interface and Information Server Design. *In* Message Handling Systems and Distributed Applications. Einar Stefferud, Ole J. Jacobsen, and Pietro Schicker, eds. Pp. 427– 445. The Hague: North-Holland.
Black, Mary, and Duane G. Metzger
   1969   Ethnographic Description and the Study of Law. *In* Cognitive Anthropology. Stephen A. Tyler, ed. Pp. 137–165. New York: Holt, Rinehart, and Winston.
Crane, Gregory
   1991   Composing Culture: The Authority of the Electronic Text. Current Anthropology 32(3):293–311.
Digital Equipment Corporation
   1986   Guide to Vaxnotes. Maynard, MA: Digital Equipment Corporation.
Feenberg, Andrew
   1988   The Planetary Classroom: International Applications of Advanced Communications to Education. *In* Message Handling Systems and Distributed Applications. Einar Stefferud, Ole J. Jacobsen, and Pietro Schicker, eds. Pp. 511–522. The Hague: North-Holland.
Finnegan, Ruth
   1991   Comment to Crane's "Composing Culture." Current Anthropology 32(3):303.
Frake, Charles O.
   1969   The Ethnographic Study of Cognitive Systems. *In* Cognitive Anthropology. Stephen A. Tyler, ed. Pp. 28–41. New York: Hold, Rinehart, and Winston.

Hiltz, Starr Roxanne, and Murray Turoff
  1993[1978] The Network Nation: Human Communication via Computer. Cambridge, MA:
    The MIT Press.
Houser, E., and R. Wallace
  1989 PC-Write Lite User's Guide. Seattle, WA: Quicksoft.
Kay, Paul, and Duane Metzger
  1973 On Ethnographic Method. *In* Drinking Patterns in Highland Chiapas: A Teamwork
    Approach to the Study of Semantics Through Ethnography. Henning Siverts, ed. Pp.
    17–34. Bergen, Norway: Universitetsforlaget.
Kerr, Elaine B., and Starr Roxanne Hiltz
  1982 Computer-mediated Communication Systems. New York: Academic Press.
Landow, George P.
  1992 Hypertext: The Convergence of Contemporary Critical Theory and Technology.
    Baltimore, MD: Johns Hopkins University Press.
Larson, Neil
  1989 Hyplus. Berkeley, CA: MaxThink.
  1990 HyperRez: An Easy to Use Memory-resident Shareware Hypertext Program.
    Berkeley, CA: MaxThink.
Metzger, Duane G.
  1988 Hypertext: The BESTNet Vaxnotes Intercultural Conference. BESTNet Report No. 5.
  1989 A Hypertext of the CSU-LA Student Ethnographies. BESTNet Report No. 15.
  1990a Hypertext Introduction and Index to the BESTNet UCR Vaxnotes Zonamaya
    Conference, Version 1.0. BESTNet Report No. 33.
  1990b Hypertext Introduction and Index to the BESTNet UCR Vaxnotes Zonamaya
    Conference, Version 1.1. BESTNet Report No. 34.
  1990c Hypertext: School Attendance Review Board Policy and Procedure. BESTNet
    Report No. 38.
Metzger, Duane G., and Guadalupe Beltran
  1990 Hypertext of Reference Classification and Bibliography of Sixteenth-century CNS
    Diffusion: Tobacco and Coca. BESTNet Report No. 35.
Metzger, D., N. Bigger, D. Díaz, J. Eckert, V. Oliva, E. Peña, and I. Polanco
  1990 Hypertext: Student Ethnographies of Colonia Xicotencatl Leyva. BESTNet Report
    No. 36.
Metzger, Duane G., and Karol Ortiz
  1989 Ethnographic Hypertext 3.0. *In* National Collegiate Software Clearinghouse, 5.
    Douglas R. White, ed. Dubuque, IA: William C. Brown Publishers.
Metzger, Duane G., and Audrey E. Weiss
  1989 A Hypertext of Marshallese Ethnographic Field Notes. BESTNet Report No. 18.
Metzger, Duane G., and Gerald Williams
  1963 A Formal Ethnographic Analysis of Tenejapa Ladino Weddings. American Anthro-
    pologist 65:1076–1101.
Nyce, James N., and Paul Kahn, eds.
  1992 From Memex to Hypertext: Vannevar Bush and the Mind's Machine. Boston, MA:
    Academic Press.
Poster, Mark
  1990 The Mode of Information: Poststructuralism and Social Context. Chicago, IL: Uni-
    versity of Chicago Press.
Romney, A. Kimball, and Susan C. Weller
  1984 Predicting Informant Accuracy from Patterns of Recall Among Individuals. Social
    Networks 6:59–77.
  1989 Quantitative Models: Science and Cumulative Knowledge. Journal of Quantitative
    Anthropology 1:153–223.
Truex, Gregory F., and Douglas R. White
  1988 Anthropology and Computing: The Challenges of the 1990s. SSCORE: Social
    Science Computer Review 6:481–498.
Tyler, Stephen A.
  1969 Introduction. *In* Cognitive Anthropology. Stephen A. Tyler, ed. Pp. 1–23. New York:
    Holt, Rinehart, and Winston.
Weiss, Audrey E., and Duane Metzger
  1990 Hypertext of the Conference CETYS:Psicologia. BESTNet Report No. 39.

von Glascoe, C., and D. Metzger
    1979 Game Cognition and Game Preference in the Yucutan. *In* Forms of Play of Native
        North Americans. Edward Norbeck and Claire R. Farrer, eds. Pp. 249–265. New York:
        West Publishing Company.
Weller, Susan C., and A. Kimball Romney
    1988 Systematic Data Collection. Qualitative Research Methods, Volume 10. Newbury
        Park, CA: Sage Publications.
Zimmer, J.
    1988 Computer Conferencing: A Medium for Facilitation Interaction in Distance Educa-
        tion. Education Media and Technology Yearbook 14.

Distance Education in Anthropology: Telecourses as a Teaching Strategy

Edwin S. Segal

Distance education can be any one of a variety of combinations of educational technologies and methodologies. The common feature among them is that they are all directed toward students who do at least part of their learning outside of the formal classroom and at some distance from any formal setting. Their surface similarities are manifestations of a deeper set of assumptions about the nature of the relationships among students, teacher, and material. In comparison with the traditional classroom and its inhabitants, students are assumed to be more highly motivated and skilled as learners; they are assumed to have less need of a teacher's guidance; and the material presented is assumed to be self-evident upon initial presentation. The homework of the traditional classroom is transformed into the major learning experience, and encounters with the teacher become an occasional experience. All of this, of course, has a profound effect on the social organization of the classroom. It is reasonable to hypothesize that the combination of a nonnormative learning setting with new approaches to the use of familiar technologies may well create either unanticipated or unrecognized learning problems for the students and similar pedagogical problems for the teachers.

Students engaged in distance education tend to be what has come to be called "nontraditional." They are older, are trying to combine college with family life, and frequently are concerned with arranging a learning schedule around full-time employment (cf. Livieratos 1989). For many of these students, separation from any formal educational environment is complete, in which case there is a strong parallel to the correspondence student. For others, there is minimal formal contact with course teachers, sometimes referred to as facilitators. This article deals primarily with students falling into the latter category.

Telecourses

One such combination of technology and methodology, though at a distinctly low-tech end of the continuum, is telecourses. These are noncomputerized, relatively simple combinations of specially developed television programs and occasional class meetings. For a number of reasons, when applied to a university setting, such distance education often seems to be no more than a device for reaching more students with fewer faculty. While this may be so, it is also true that distance education through a telecourse medium has its own characteristics and may even have its own strong points. This article assesses the educational impact of a telecourse designed to present an introduction to cultural anthropology.

Part of this article stems from the results of two comparisons of student performances in introductory courses (Segal 1986, 1990). Because in each instance one of the two courses was a day section and the other was an evening section, the comparison affords an opportunity for examining the extent to which differences in performance can be ascribed to different student populations or different teaching media. Although the following course descriptions refer to courses taught in 1990, the pattern is closely paralleled by the earlier exploration.

Course Profiles

Typically, a 200-level introductory course offered during the day draws an initial enrollment of about 100. This course was somewhat smaller, beginning with 78. Typically, daytime introductory courses exhibit a steady decline in enrollment until the formal cutoff date; and, thereafter, de facto dropouts continue the trend. This daytime course was no exception. By the end of the semester, enrollment had dropped to 63. Evening introductory courses usually have an initial enrollment between 15 and 25. This course was considerably larger, having an initial enrollment of 45. In fact, the size cutoff was produced solely by room capacity. By the end of the semester, enrollment had dropped to 33, reflecting the typical daytime pattern. More usually, evening courses have a relatively stable class size.

Course Organizations

The ordinary introductory course was a day section, meeting three days per week, with an initial registration of 78; the telecourse was an evening section. The text for both courses was the one to which the telecourse was keyed (Haviland 1990). The day course was organized in a standard lecture/discussion format, with occasional use of audiovisual materials; it was taught on the university's main downtown campus, and utilized, as a secondary optional text, a student workbook and study guide designed to accompany the text.

The evening course, advertised by the university as a more convenient way of going to college, was organized around the broadcast schedule of the local Public Broadcasting System affiliates, and was taught on the university's suburban campus. The Public Broadcast facilities in Louisville are such that it was possible for each broadcast to be repeated five days per week, at varying times. In addition, the class met, as a unit, eight times during the semester for two hours each meeting. Each class session was devoted to lecture/discussion and, as warranted, exams. There were a total of 13 broadcasts. Each hour-long broadcast consisted of two half-hour segments that followed the course outline provided by the Haviland text and associated course materials distributed by Coast Community College. The class also used a student workbook/study guide designed to fit the materials as presented in the television broadcasts.

In an effort to avoid the potential for passive viewing, students in the telecourse were asked to keep journals in which they answered questions related to the broadcasts. Journals were not graded but were collected periodically. Students were told that there would be no journal grade and that the sole purpose of the assignment was to encourage them to be active television viewers.

Course Comparisons

Evaluations of teaching effectiveness focus most often on mastery of course content as measured by examination performance. In terms of an approach to teaching, this constitutes a separation among different forms of knowledge: cognitive, attitudinal, and affective. Cognitive knowledge primarily refers to those items that can be called "facts," either cultural or supposedly objective. These are items that are more or less easily tested on multiple-choice exams. Most students in introductory courses of any format also seem to think that the course and the exams are largely concerned with this type of knowledge. Cognitive knowledge constitutes the core material for the more and more fashionable performance criteria.

The distinction between attitudinal and affective is a bit fuzzier but, nonetheless, important. Affective domain knowledge refers to aspects of approaches toward learning grounded in emotional, rather than cognitive, responses to the material presented. Most commonly, the affective domain is seen as being associated with student motivation. Elements within this motivational subdomain include techniques for maintaining the student's attention, helping the student perceive relevance, helping the student gain confidence, and helping the student feel satisfied with the educational experience (Zvacek 1991). Affective domain elements are important in the ultimate success of an educational endeavor. It is also reasonable to expect a difference here between telecourse and "traditional" students.

The attitudinal domain of knowledge is not unique to teaching anthropology, but it is important to that enterprise. Here, I refer to concepts and conceptual categories (e.g., critical thinking), analysis (as opposed to opinion), cultural relativism (as opposed to primitive and civilized), theory, evolution, race, and prejudice. Labeling is easy, as is raw definition (e.g., a theory is), but direct measures of whether or not students have learned to think in a more analytical way, or a more culturally relative one are difficult to construct. However, the distinction is especially important to anthropology because teaching new cognitions, exotic facts about our own and other cultures is only a part (and for some, frequently the smallest part) of our educational agenda. This aspect of the comparison is examined from material gathered through the journals and casual conversations with students.

The two classes studied in 1990 were each given four identical multiple-choice examinations, three during the semester and the final. An initial comparison of means and standard deviations for exam scores for each class reveals an interesting pattern (Table 1).

Table 1
Means and Standard Deviations for the Four Exams

	Traditional Course		Telecourse	
	Mean	SD	Mean	SD
Exam I	76.9	9.98	75.4	13.96
Exam II	73.9	9.88	74.9	11.27
Exam III	68.9	11.17	72.0	9.62
Exam IV	66.9	10.03	71.8	7.38

Both classes begin on a relatively even footing. Although the evening telecourse class has a greater dispersion around the mean, at $p < .05$, it is not significant. In fact, as Table 2 shows, none of the differences are significant until we reach the final exam. These data seem to be a clear example of growing differences over the course of the semester, suggesting that we truly need to rethink performance evaluation and approach performance as a process.

An earlier examination of the same problem (Segal 1986) produced somewhat different results. In the earlier instance, the only examination scores showing a significant difference were those on the first exam; the standard deviations were more stable but tended to decrease slightly in the day class. At that point, one reasonable conclusion seemed to be that in the final analysis the telecourse taught cognitive material as well as the more traditional classroom presentation. This is a conclusion in line with other studies of other telecourses (Blanchard 1989). Although this set of data suggests additional interpretations, it does not contradict this essential finding. The question really becomes one of whether or not we, as teachers of anthropology, are willing to settle for simply teaching "the facts."

Table 2
Statistical Comparisons for the Four Exams

	Difference of Means	Analysis of Variance
Exam I	$t = .585, \ df = 107$	$F = .751, p = .608$
Exam II	$t = -.413, \ df = 104$	$F = .611, p = .558$
Exam III	$t = -1.322, df = 100$	$F = 2.412, p = .119$
Exam IV	$t = -2.607, df = 96$	$F = 6.539, p = .012$

Casual conversations with the students indicated that the telecourse concept had drawn many of them to enroll in anthropology, rather than a different social science. To the extent that the students' initial motivation is a significant aspect of the learning environment, the telecourse, simply by being, may have an impact on what is learned and how well.

Student Profiles

The telecourse students in both the 1990 and 1986 classes closely fit those characteristics reported for telecourse students in general (cf. Livieratos 1989; Washington State Board for Community College Education 1990). A brief summary of the characteristics of the 1990 groups is presented below in Table 3. Virtually all students in both classes also reported that they are employed in a wage-earning capacity, which is typical of telecourse students and increasingly of the more traditional student population.

The popular wisdom is that evening students are more "fun" to teach. They are generally older, have a larger fund of experience to bring to the class, are more interested in learning, and tend to make the discussion aspect of a lecture/ discussion format more lively. On the other hand, the lack of extensive direct student-teacher interaction under the telecourse structure tends to reduce students' comfort with discussion, even while they are exposed to larger quantities of material between meetings.

Differences in course performance can be tested easily by examining the set of intercorrelations among the variables listed in Table 3 and the four exam scores. These were calculated first for both classes as a single sample, and then for each class separately. Women were coded 1 and men 2;

Table 3
Background Data

	Day course	SD	Telecourse	SD
Mean Age	21.35	4.59	32	10.85
Female	39.7%		60.6%	
Mean GPA (4-point scale)	2.57	.79	2.98	.7
Freshman	30%		9%	
Sophomore	28.5%		21.2%	
Junior	28.5%		45.5%	
Senior	9.5%		18%	
Other	3%		6%	
Mean TV Viewing (hrs./wk.)	4.6	2.2	4.52	2.39
Mean Course Study (hrs./wk.)	2.81	1.9	3.64	1.98
Mean Final-Exam Study (hrs.)	7.23	7.43	8.62	6.28
Used the Study Guide	55.6%		69.7%	

Table 4
The Intercorrelations[a]

Name	Total Sample	Day Course	Evening Course
Gender—Course Study	−.146	.038	−.384*
Gender—Study Guide	.376*	.334*	.413*
Age—Study Guide	−.218*	−.096	−.273
Age—Class	.378*	.381*	.258
Age—GPA	.023	.219	−.338*
Course Study—Exam Study	.523*	.505*	.545*
Course Study—Study Guide	−.287*	−.246	−.312
Exam Study—Study Guide	−.271*	−.322*	−.108
Class—GPA	.089	.267*	−.307
GPA—Exam I	.528*	.626*	.395*
GPA—Exam II	.551*	.594*	.499*
GPA—Exam III	.451*	.572*	.257
Age—Exam IV	.221*	.220	.018
GPA—Exam IV	.641*	.732*	.469*
Exam I—Exam II	.714*	.645*	.811*
Exam I—Exam III	.608*	.585*	.665*
Exam I—Exam IV	.683*	.732*	.566*
Exam II—Exam III	.546*	.440*	.783*
Exam II—Exam IV	.679*	.666*	.712*
Exam III—Exam IV	.789*	.811*	.739*

[a] Only correlations significant at $p < .05$ in one of the three samples are reported here.   * = significant at $p < .05$.

using the study guide was coded 1 and not using it was coded 2; class is ordinal and all other variables are interval scale. Negative correlations with gender reflect a tendency for women, rather than men, to be associated with the variable; negative correlations with the study guide reflect a tendency for its use to be associated with the variable.

The extent of similarity in the two samples is striking and supports Blanchard's (1989) generalization that the telecourse structure teaches cognitive materials as effectively as the more traditional classroom lecture/discussion structure. To the extent that performance on early exams is a strong predictor of performance on later exams, we can certainly say that both classes exhibited progressive learning throughout the semester. In the same vein, the best predictor of success on the first two examinations seems to be total previous performance as measured by self-reported grade point average (GPA). It is heartening to observe that in both classes some measure of previous performance is a strong predictor of subsequent performance. A multiple regression approach to a more detailed analysis does not challenge the conclusions. The analysis only adds weight to the extent to which age and class standing can be seen as effective precursors of GPA. Essentially, more experienced students have done better and continue to do so, not a very startling result.

Students were originally asked to estimate the number of hours per week, aside from any time associated with class work, that they watched television. This was meant as an approach to the hypothesis that some level of visual, and especially television-oriented, literacy would be of benefit to students in the telecourse. The absence of any significant relationship between television hours watched and course performance suggests that visual literacy, as measured here, is not an important aspect of the effectiveness of the telecourse. On the other hand, there are some indications that the visual materials of the telecourse are problematic for some students and that assumptions about "self-evidentness" may be unwarranted.

Discussion

A major difficulty with many studies of teaching techniques lies in their experimental construction. To the extent that students in an experimental class have been placed there in some way other than voluntarily, the motivation for classroom performance is either absent or distorted (Goldsmid and Wilson 1980). To that extent, the nonrandom origins of the study populations have a virtue in that all students in both classes were there for the same variety of reasons. This is but one aspect of the general "field test" orientation of this study. All variables were not strictly controlled, samples were not carefully drawn to be representative of some abstract population. In effect, two ordinarily occurring classes in cultural anthropology were examined within their sociocultural educational matrices.

The conclusion that the telecourse structure seems to be at least as effective as a more traditional approach does not, at first, seem very different from that of Dubin and Taveggia (1968) that the array of devices used in teaching is irrelevant to outcome. But in this instance that conclusion is actually very different from theirs. The ordinary day course occupied 16 weeks with two-and-a-half class hours per week, a total of 40 hours of instruction. The telecourse consisted of 13-hour-long broadcasts and eight two-hour class meetings, a total of 29 hours of instruction. At least in terms of cognitive learning, the 29-hour telecourse system is as good as the longer 40-hour standard course. Essentially, this means that if convenience in education means spending less time in absorbing enough facts to pass a series of examinations, then the university was quite right in advertising "Faces of Culture" as a more convenient way of going to college. On the other hand, this should not be taken to imply that a telecourse utilizing 40 hours of instruction would necessarily teach any more material.

However, there is still a large number of potentially significant, quantifiable factors not addressed by this study. For example, because the programs were each broadcast five days per week, we have no way of knowing whether students in the telecourse watched each one more than once, equivalent to attending the same lecture more than once. Another aspect involves ownership, or access to a VCR, the technological ability to make the broadcast schedule even more convenient, or to make the programs even more repeatable. I do know that the day-class students did not watch the programs. During the course of lectures each week, I managed to simply ask if anyone had seen a particular item on television. No one ever said yes.

In the same vein, but on the other side, just as approximately 20 percent of students in the day course regularly cut lectures and rely on textbook reading and "getting notes from someone else" to get through, we have no measure of how many telecourse students "cut" programs and relied on text reading and getting notes. The journal assignment was faithfully carried out by all students for all program segments, indicating that in one way or another, they had all gained at least some minimal familiarity with the material, but the source of this familiarity is unknown.

One question that plagues all educational endeavors—retention beyond the final exam—is not really addressed here. This is clearly beyond the scope of an article such as this one, but I would like to note the number of times someone has said to me, in casual conversation, "I had an anthropology course once, and all I remember is the fascinating . . . [fill in your own ethnic group name]."

The questions of attitudinal and affective education are especially difficult in a time of increasing enrollments and decreasing budgets that create continuously growing pressures for finding more ways to attract more students, and more ways to handle more of them at once, with fewer resources. Although most of the studies cited here report high levels of student satisfaction, this seems to be based much more on the convenience with which major portions of the requisite materials are delivered, than it is on learning a great deal, or on learning new ways of viewing the world. Johnson and van Iten (1984) found that at least in the area of language learning, telecourse students did not achieve the levels of proficiency achieved by students in more traditionally oriented language classes. To the extent that learning a natural language parallels learning a disciplinary language, we may have here an indication that some things important to the teaching of anthropology may not be being learned as well as we would wish.

Telecourses and telecourse materials also raise a related issue: one of academic quality and considerations of the biases and distortions built into the materials being used and of the pedagogical goals involved in teaching an introductory course. These questions are, of course, present in any teaching situation. However, because the telecourse context diminishes the extent of a student's opportunity to actively interact with any of the educational media being used, the issues are sharpened. From the instructor's point of view, it was someone else's agenda informing the construction of the televised materials. The instructor was consulted about neither the materials being presented, nor the juxtapositions of segments, nor the text of any narration.

For "Faces of Culture," I found little on the surface to complain about in most of the narration, and the visual portion seems to have culled through most of the extant excellent ethnographic footage. On the other hand, the series has come under criticism for presenting a "colonized anthropology" whose use of the ethnographic present and out-of-context film clips regularly produces background images of primitive peoples doing primitive things, while the foreground narration talks about cultural relativism (Buck 1989). At the same time, I found myself in strong disagreement with what I thought were basic theoretical issues in the Haviland text, and regularly taught "against" the ethnocentric interpretations my students (in both classes) were prone to make on first response. We have no true measure of the relative contributions to the learning process of narration and visual image, neither do we have similar measures for text, study guide, or program as a whole.

Student discussions, responses in journals, and casual conversations do highlight some differences in what can be called the culture of the classroom. The occasional use of audiovisual materials in the ordinary lecture/discussion class does not seem to leave students with the sense that they are studying real people living today. On the other hand, although Buck may be right about the effects of the ethnographic present on student perceptions, telecourse students are also more likely to think of the Nuba, Aymara, or Yanomamo as real groups of living human beings. Telecourse students also seem to be more likely, in discussion, to draw parallels with their own experiences. However, this may be as much a function of greater age and greater life experience as it is of the telecourse.

Telecourse students are also more likely to ask questions or to start discussions, but the source of this difference is not necessarily encouraging (in either direction). The more traditional students are likely to say things like: "The lectures seem so clear when you're talking. . . ," while the telecourse students are likely to say things like: "Among the many things I didn't understand in the programs we watched. . . ."

Finally, there is the student who commented on the whole telecourse with:

> This is probably the hardest course I've ever taken. The books don't provide much in the way of clues for what to watch, and we don't get to meet with you often enough. It's a real crapshoot out there figuring out what's important and what's not.

This comment came from a student who had been given prepared questions for each of the programs and who ultimately received a grade of B. This is not the only such comment I've received, but it is the most articulate. Her major point was that in spite of all the effort I put into preparing guidance ahead of time, when it came time to watch a program, she needed some equivalent to the textbook's underline, bold-face type, or col-

Table 5
Evaluative Responses from Telecourse Students

Liked Most	N	Liked Least	N
The television programs	4	The television programs	3
Infrequent meetings	2	The meeting schedule (too infrequently)	3
Self-paced study	3		
The class discussions	5		
No response to either question: 12			

ored box. If a relatively good student finds little guidance in the programs themselves, then what are other students missing?

Another aspect of the impact of the various parts of the course can be seen in student responses to two questions: "What did you like most about this class?" and "What did you like least about this class?" (see Table 5). Out of 25 respondents, three cited the programs as the least desirable aspect and three cited the every other week meeting schedule. Four thought the programs were the best aspect, two pointed to the meeting frequency, three cited the organization around self-paced study and five thought the class meetings were the best part of the course. Almost 50 percent of the class did not bother to mention the television programs as either a high or low point of the class structure.

Several students commented that the best way to improve the course was to have more class meetings. In that context, several students complained that the class meetings, as scheduled, did not provide adequate time for discussing the material in either the video programs or the text. The audiovisual presentation is not as self-evident as seems to be assumed. Essentially, the teacher still has considerable teaching to do and less time in which to do it. While students are more in control of their learning schedules, they are no more in control of the ultimate outcomes.

Contemporary technologies and teaching/learning packages are frequently presented as a wave of the future. In reality, there is considerable cultural continuity between high school norms and pedagogical paradigms in college. To the extent that secondary education utilizes an empty vessel paradigm, major portions of a telecourse are continuous with it, even while new cognitive material is being presented. This is especially noteworthy because so many of the telecourse students are not drawn from the traditional population pools. Telecourse students learn definitions of culture or ethnocentrism but do not seem to have much opportunity for asking the questions or engaging in the discussions that apply the concepts.

Future examinations of telecourses need to pay closer attention to measures of affective and attitudinal learning. One approach to is treat examination questions as an inventory of informant interviews. One variety or another of cluster analysis can then be interpreted as providing a picture of "cultural consensus" within the class. Carried out over several semesters, a comparison of first exam and final exam would begin to give us a sense of the extent to which this consensus changes under telecourse or other conditions. Additional attention also needs to be focused on exploring questions of visual literacy. Short student essays describing how they approach the television programs would provide the kind of qualitative data necessary. Asking students to keep a simple log of when they study, or watch the television programs, and for how long would provide some of the quantitative data necessary for evaluation of telecourse effectiveness.

Even with the incomplete studies reported here, it is quite clear that telecourses formatted as film clips with voice-over narration need some basic format redesign. In the "Faces of Culture" series, the narrator seems to have been chosen for his very cool and uninvolved vocal qualities. For many students, this proves to be much more soporific than objective. The narration itself needs to provide a better sense of what the important points are—perhaps including a brief summary, in the middle and at the end of a program highlighting important points, without the distraction of visual images. Finally, perhaps the use of different narrators for different programs will provide some of the apparently necessary variety. In this regard, it would also help if some of the narrators were women, and if all were also members of a variety of ethnic groups, or possibly nationalities. In the final analysis, we need to keep in mind that the goal of a telecourse is supposed to be effective education, not simple entertainment.

Possible Implications

If we think of the telecourse as one model for using video materials in a classroom setting, this study also lends itself to a bit of useful speculation about the model's effectiveness. If we think of a university classroom as a species of extended training session, then it seems reasonable to ask about the use of this teaching model in other similar longer or shorter training sessions. Video presentations are efficient mechanisms for delivering data in relatively short periods of time. People more experienced in the field for which they are being trained will probably get more out of video presentations. But regardless of experience, video presentations are probably not very useful without an accompanying discussion. They simply do not seem to replace teaching beyond a short-term expansion of cognitions. In this regard, it is important to remember that the extent of long-term retention is virtually unstudied.

One area in which this extension of these data has potential significance has to do with various kinds of multicultural training situations. Whether we are trying to teach students to come to a better understanding and acceptance of their diverse campuses, trying to bring workers in a factory to a similar attitude toward their work places, or trying to ease tensions in a changing neighborhood, the video approach seems most capable of presenting some data and stating some arguments. Without a human teacher/facilitator/group leader to draw out what the video is intended to convey, it probably will not have much effect.

In the end, this technological device, and likely most others as well, does not seem capable of replacing extensive student-teacher interaction. The implication is that as we move into a world in which we try to do more with fewer resources, we need to rethink our educational models and focus on enhancing teacher effectiveness rather than helping a teacher deal with more students at one sitting. Classroom consolidation probably has an easily reached point of diminishing returns.

Notes

An earlier version of this article was presented at the Annual Meeting of the American Anthropological Association, New Orleans, LA, November 28 to December 2, 1990.

References Cited

Blanchard, William
    1989 Telecourse Effectiveness: A Research-Review Update. Olympia, WA: Washington State Board for Community College Education.
Buck, Pem
    1989 Review of "Faces of Culture." Notes from the ABA 15:1.
Dubin, Robert, and Thomas C. Taveggia
    1968 The Teaching-Learning Paradox: A Comparative Study of College Teaching Methods. Eugene, OR: Center for the Advanced Study of Educational Administration, University of Oregon.
Goldsmid, Charles A., and Everett K. Wilson
    1980 Passing on Sociology: The Teaching of a Discipline. Washington, DC: American Sociological Association/Teaching Resources Center.
Haviland, William A.
    1990 Cultural Anthropology, 6th ed. New York: Holt, Rinehart, and Winston.
Johnson, Margaret S., and Helga B. van Iten
    1984 An Attempt at Televised Foreign Language Instruction. ADLF Bulletin 16(1):35–38.
Livieratos, Barbara B.
    1989 Distance Learners: Howard Community College's Fiscal Year 1989 Telecourse Students. Research Report, 62. Office of Research and Planning, Howard Community College, Columbia, MD.
Segal, Edwin S.
    1986 Do Telecourses Teach Anthropology? Paper presented at the Annual Meeting of the American Anthropological Association, Philadelphia, PA, December 3–7.
    1990 Teaching a Telecourse. Paper presented at the Annual Meeting of the American Anthropological Association, New Orleans, LA, November 28–December 6.
Washington State Board for Community College Education
    1990 Video Telecommunications in Washington Community Colleges: A Descriptive Study of the Current Activities and Future Plans. Operations Report, 90-1. Olympia, WA: Washington State Board for Community College Education.
Zvacek, Susan M.
    1991 Effective Affective Design for Distance Education. Tech Trends 36(1):40–43.

The Interpenetration of Technology and Institution: An Assessment of an Educational Computer Conferencing System

James H. McDonald

Introduction

I'm lost. Help! Is there anyone on the other side? If someone's out there, please answer back. [Student entry on the CoSy system, Spring 1990]

I felt like the Wizard of Oz as I read this student's plea for help. Unfortunately, I was not in the Land of Oz, and this was not a message from Kansas. Rather, I was an instructor on "CoSy," short for (Co)nferencing (Sy)stem. CoSy[1] is an interactive electronic bulletin board employed by an Arizona community college as an alternative method of course delivery.

The notion of increasing an individual's access to education was the administration's official goal for CoSy. The computer was treated by the administration as a new, innovative way to create interactive courses for people who would otherwise not receive an education. The community college recognized the need to serve people who cannot fit into a traditional classroom setting with its rigid, fixed schedule.[2] The use of computer networks and other alternative delivery methods provided an institutional solution to meet the needs of the truly nontraditional student, particularly those individuals re-entering the educational system. In practice, however, drop-out rates in CoSy classes were high, throwing into question the success of the program.

CoSy has three modes of communication: conference, mail, and converse. Conferences and conversations are public areas where a student can read messages left by others, reply to them, and add new messages. Conferences might best be thought of as the classroom area in CoSy. Conversations are set up by instructors who may include all or part of his or her student population. Conversations are commonly used to break larger groups into smaller discussion or work groups. The mail system allows private electronic messages to be sent to one or more named individuals. The terms used to describe the different functions within CoSy seem understandable because the meanings require little decoding or specialized computer language knowledge. Despite this, the student's cry for help and the superficial accessibility of the terms used in CoSy should be taken as a first warning of a gap between appearance and substance, presupposition and reality.

The feeling of being lost is highly alienating.[3] The student was expressing her lack of a coherent sense of the CoSy network and the tasks at hand. Her experience, furthermore, was far from rare. She was in the company of many other students who were confronting a computer or other form of technology for the first time (cf. Dreyfus and Dreyfus 1986; Walters 1988). They did not have the conceptual knowledge necessary to navigate through CoSy or any other unknown computer application.

The use of computers in education is part of a broader emphasis in our society on "computer literacy" (Noble 1984). The rationale for computer literacy is often framed in survivalist terms—do not get left behind, be competitive! We are currently flooded with computer advertisements in the media that inflate these ambiguously defined and generally ill-conceived fears.

As a consequence, scores of parents are rushing out to buy computers to ensure their children's future marketplace competitiveness. Many of those worried parents, as well as ourselves, carry certain presuppositions about the meaning of computer literacy. But what is it? What are the implications of that definition? Who defines its meaning? Does passing knowledge of WordPerfect 5.1, for example, indicate literacy?

When I began teaching over the CoSy computer network, I was abruptly confronted by my own computer illiteracy, even though I managed to spend most of my waking hours in front of a computer. I had never used telecommunications in the past. Like most of my students, I had to learn the basics of the system, while at the same time going about the business of my course. It was expected that students and instructor alike would "pick it up on their own."

Not surprisingly, under these conditions there was a high drop-out rate for the CoSy-based courses. Each semester between 1989 and 1991, we lost approximately 50 percent of our student population. While some of this was attributable to simple problems of scheduling and logistics, it was critical to ask about the connections between alienation and attrition on the one hand, and the community college's perception of student identities and the meaning and role of the machine on the other.

This article explores the computer network as a type of "black box." Latour (1987:2–3) suggests that we create conceptual black boxes when "a piece of machinery or a set of commands is too complex. In its place we draw a little black box about which [a person] needs to know nothing but its input and output." Computers are one of those black boxes. We take them for granted and speak of them as "tools," objects so much a part of our daily lives that we treat them as if they were neutral and natural parts of the landscape. I am concerned, conversely, that students become familiar with the cultural logic behind computer design and application: what is inside the CoSy black box?

By black box, I mean more than just the technology, but also the field of social relations that emerge around the machine. Administrators, instructors, students, and the machine are all linked together, forming a system. It was the community college administration that had the most powerful hand at defining the nature of the CoSy black box because they controlled all aspects of CoSy delivery. While they did not design the software or hardware, they chose what to use and how to apply it to education. The relationship between bureaucracy and technology was captured, for example, in a computer designer's critique of a competitor's machine:

> He decided that the VAX embodied flaws in DEC's [Digital Equipment Corporation] corporate organization. The machine expressed the phenomenally successful company's cautious, bureaucratic style. [Kidder 1981:36]

It is pertinent to ask then, what does the configuration of the CoSy system reflect about the nature of the community college bureaucracy? What institutional, bureaucratic logic has been constructed around the use of computer-based education that both facilitates and hinders student success?

In sum, I have posed a series of questions: what is computer literacy; why are students alienated; and how has bureaucratic logic influenced these outcomes? It had created an exclusionary social field, permitting some students to access an education using the CoSy system while effectively denying this opportunity to others. My concern is with how access was gained or denied, and as a result, who was likely to be included or excluded from using CoSy successfully. At issue in this article is how the design and administration of CoSy resulted in the differential access to education by more and less advantaged groups,[4] and how this process might be remediated. Implicitly, then, this analysis is also about power, social control, and the production and reproduction of bureaucracies.

Research Methods

Research at the community college began with formal surveys of students taking introductory anthropology courses on the CoSy system. From the Spring of 1989 through the Fall of 1991 I collected 93 surveys concerning student experiences with computer conferencing. I also conducted informal interviews with numerous students, and several of those individuals functioned as "key informants" providing information on their experiences with CoSy and the community college administration.

In the Fall of 1990 I began to work for the community college as a student and faculty liaison. In addition to teaching on the system, I also developed documentation for CoSy, conducted training sessions, and actively troubleshot technical problems confronted by students and faculty throughout the semester. An important part of my duties involved attending meetings, discussing marketing, and creating course delivery strategies with key administrative personnel at the college. This change in status greatly broadened my contact with students, faculty, and administrators involved with CoSy. The position necessarily involved an intimate knowledge of the everyday operation of CoSy through student and faculty interaction but also provided a window onto administrative logic and decision-making processes.

Culture and Technology

In assessing how cultural logic is inscribed on technology, I have found it useful to employ the concept of *format* that assumes technology is a cultural construct that may take many different forms (Altheide 1985a, 1985b, 1989). The computer is certainly not a passive vehicle for the distribution of educational resources (see Bowers 1988). It is, therefore, critical to examine the different vectors of its formatting (i.e., ensembles of procedures, instruments, actors, and discourses and the kinds of relationships that are structured—fields of relations and what can and cannot be done within them) (Altheide 1989:189). A format defines limitations in time and space of the types of social relationships and forms of communication that can take place among people. A particular format, furthermore, can produce different outcomes across different populations. Technology should not, consequently, be thought of as an object, a tool. Rather, it shapes the social field of action, making some types of behavior possible and others less possible or altogether impossible (Wolf 1990:587).

Seen as a cultural construction, *the computer is a complex sign upon which multiple meanings are inscribed.* Technology is at once an apparatus, a discourse (about progress, efficiency, control, power, jobs), and the relationships they create between people, and between people and machine. Technology, from this perspective does not just interface with the institution, it *interpenetrates* the institution and results in a new social form. Computer technology also presupposes a certain user identity. This is most clearly etched in the software chosen for the computer, as well as in hardware, such as the keyboard (Altheide 1985b). It is also contained in certain institutional features, including home-based versus lab site access to course materials and forms of student support (e.g., orientations, CoSy users' guides, technical advice). All of these factors together constitute the identity of the system.

Altheide (1989:187) emphasizes how technology, and implicitly, bureaucracy, define a social field and the boundaries of possible action. In addition, I will stress the potential of the people within those fields to create various forms of resistance to and redefinition of the social field. The technology, the nature of the organization deploying the technology, and the users of that technology all contribute to the shape and outcome of an innovation, such as the introduction of computer conferencing in a community college. All three elements are intimately intertwined and result in a context-specific outcome that cannot be predicted from the outset (Barley 1986; Dubinskas, this volume).

Student Experiences on CoSy

The students enrolling in CoSy-based courses were generally lay us-ers of computers who lacked an understanding of the technically sophisti-cated world of computers and did not particularly care about computers. The same could be said of all of the instructors using CoSy. We were tech-nological "fringe dwellers." We were not particularly interested in the tech-nology for technology's sake, but rather saw it as a means to an end. In my case, I hoped to increase course availability for students who could attend traditional classes. For the majority of my students, the goal was to com-plete a course requirement in a way that was not only convenient, but was often their only realistic alternative.

The system, however, had its shortcomings. Exposure to course mate-rials was different for CoSy students than for students in the traditional classroom. In a classroom, students share a *common space, hear* the lec-ture, *write* down notes, and then can *visually* review those notes. A student can access the course information using three different modalities. Further-more, a lecture is a kind of performance with verbal and nonverbal cues about the relative importance of information and ideas. There are also jok-ing asides, spontaneous commentary, and interactions that give each class a unique chemistry.[5] The holistic experience of the classroom is lost when similar information is taught over the computer. Students receive in-formation in written form only, thereby reducing the number of times and ways they process the material. Additionally, students cannot ask ques-tions with the spontaneity accessible to them in the classroom (although they can ask questions and receive responses far more quickly than through traditional print-based correspondence courses). While CoSy is highly interactive, there was still a time gap between questions and re-sponses.

Computer-based courses also place a heavy burden on the student. They must have enough self-discipline to stay current with the course. The advantages provided by the system's flexibility can be a problem for less disciplined students. It is very easy to put off work in an electronic class if the student is not immediately held responsible for his or her presence and participation. On the other hand, this problem is no greater in many ways than large lecture classes where students are essentially anonymous. However, in a classroom-based course there is at least a fixed schedule providing structure to the learning experience. In computer-based courses, the student must largely impose that structure.

Aside from the logistical difficulties inherent in distance learning, the single largest problem was the range of computer competency students brought to CoSy classes. For example, simple technical knowledge of computer skills (i.e., word processing), while necessary, was only part of the aptitude necessary for using the system. Accessing CoSy also required the use of telecommunications.[6] The successful use of CoSy, therefore, required a multifaceted knowledge.

Once connected to the CoSy system, the student quickly found the system quite user-unfriendly. CoSy was designed by programmers proficient only in computer logic and function, traceable through the system's lack of menus and other aids that would make it easier for novice use. Various features of the system were clearly designed without input from the field, and overall it compared poorly to industrial systems. A cryptic system design, therefore, was a major barrier to understanding and accessing the system. This was an even greater problem when we consider that students frequently lacked computer and telecommunications experience. The danger of this inadequacy was that students either became quickly frustrated with the system and dropped the course or they spent more time and energy learning to use the system than on actual course content. The dismay of the students was reflected in the attrition rate for CoSy-based courses. Most of the students taking these alternative courses were older, returning students were seeking to upgrade their credentials in an increasingly competitive job market. Returning students, in general, have a tremendous amount of anxiety and low self-esteem accompanying their re-entry into the educational system. Many confided that they did not feel as though they could successfully compete with their younger counterparts. This apprehension and fear carried over to computer use. Many CoSy students felt that the technology had passed them by and that they lacked the aptitude necessary to understand the machine and its operation.

Student interactions with computers generate new sets of information and meaning apart from actually learning a specific area of course content. One of the most important relationships generated between computer and student is one defining user competency (Walters 1988:18). As students tried to understand the computer, they were constantly receiving messages about their level of competency. Concurrently, they were being challenged by the actual course content. Thus, the normal demands placed on re-entry students were exacerbated by the technology and were often expressed as fear. This fear was captured by one student who said:

> I have had a lot of anxiety which I credit to my age and the fact that I have been out of school for a long period of time. I am afraid of failing, therefore I try extra hard and this brings on the anxiety attacks.

Students also expressed significant reluctance to ask questions over CoSy and engage either the instructor or their fellow class members unless coerced to do so.

> The disadvantages [of CoSy] are that when I had a problem, I hesitated to ask on-line.

Ironically, virtually all students also emphasized that they felt isolated, especially those who worked at home. A consistent theme in student commentary was the "lack of exchange of ideas" between students. Most students complained about their inability to get to know their classmates. These students had to conduct their interactions on an "electronic stage" that was decentralized and dispersed—they were alone accessing only fragmentary, "cold" information (Poster 1990:66). They were not in a classroom participating in an original, "hot" event.[7] It was difficult for students to distinguish the "real" (e.g., a class or a living, breathing instructor) from the "hyperreal" (e.g., the mediated representation of the "real"). It amazed me to learn from discussions with CoSy students that they rarely talked to their instructors over the phone, let alone meet with them in person. The class members and the instructor did not exist at an absolute point in time or space, but rather they (re)materialized on the computer screen 24-hours a day (Poster 1990:73). As one student bemoaned, "we communicate only through carefully chosen written words." Her comment expressed an understanding at some level of the self-policing that occurs when spoken language is written and is subject to the careful scrutiny of the instructor and other class participants.[8]

Conversely, students also had positive experiences and interpretations of the system. One student, for example, felt that the computer-based courses were liberating:

> I enjoy learning, but it is difficult to relate to 18 and 19 year olds, and what *they* feel are life's tragedies and expectations. This class offered [me] a chance to learn without feeling *old*. [emphasis added]

Still other students mentioned that they found traditional classroom-based courses annoying because someone was always "wasting their time" in class by asking irrelevant questions or diverting the flow of the course in some way.

> I [would] miss the experience of other students and a lecturer if I could get the guarantee that they were personalities I would enjoy. If they were unenjoyable (you know—boring, monotonous, unenthusiastic), then I don't miss the interaction at all.

Finally, computer-based technology was the only way for some students to access the courses required for a degree. One student confided,

> I was unable to get my last science course due to conflicts that occur with set times for classes and baby sitters, not to mention the reduction of pay due to my inability to work. [CoSy] allowed me the freedom to complete my needed requirements without a loss of income.

Student Resistance

Students responded to the CoSy system both passively and actively. The high drop-out rate in CoSy-based courses was the clearest form of passive resistance to the system. Students, however, also worked out a number of other active and creative resistances to the system, including attempts to redefine the social field that had developed around CoSy.

The student who noted that we communicate through "carefully chosen written words" understood that the discourse we have over the computer is limited and limiting in many ways. The computer defines how, when, and with whom we can communicate. CoSy students made a number of creative attempts at redefining the CoSy system in ways that were not originally intended, and in some instances they succeeded.

I had many students, for example, who urged face-to-face meetings to discuss their work, exams, and so on. This meant in many cases that students gave up what little free time they had, lost paid working time, and/or travelled long distances. Nevertheless, they wanted to meet in person. Meetings of this kind redefined the social field of discourse for my computer-based classes. One student referred to these meetings as, "Putting a little human experience back into the class." Yet another said, "[These meetings] gave character and depth to the words on the screen. I could just imagine [the instructor's] expressions as I read various phrases." Similarly, students phoned me just to "chat." What is interesting about this is not merely that the students initiated these actions, but that many of their other instructors implicitly defined these activities as outside of the limits of this field of class-based interaction.

Mass panic was another form of resistance generated by students. This happened commonly when the mainframe computer housed at the district office "crashed." This was usually followed by a rash of phone calls to the main office by worried students who thought their computers or modems no longer worked because they would not connect to the mainframe. Lacking the knowledge necessary to understand how the system worked and how to tell when something was wrong, these students found an outlet for their frustration and fear. They had yet to develop the experiential base to scan their own computer's functions, know that it was working properly, and then realized that the mainframe had malfunctioned. When part of their panic was connected with getting assignments to their instructor on time, they further revealed a lack of understanding that when the mainframe was down for them, it was down for everyone.

There were also resistances created by the more user-sophisticated CoSy students. I was contacted by a number of individuals who volunteered to rewrite some of the CoSy program to make it easier to use. Still others modified their communications software in numerous ways to make their interaction with CoSy work smoothly.

On a potentially destructive level, there was also some "hacking" within the CoSy system. Twice in the Fall of 1990, hackers, commonly disgruntled students, entered into CoSy and mailed threatening notes to various instructors and staff. Luckily, the hackers showed a limited knowledge of the system and a lack of any deep understanding of the technology. They created some mayhem, but they never disabled the system.

The issue of hacking raises a paradox. The more thoroughly we train individuals to understand the logic of computers, the more vulnerable a system is to hackers. Student control of computer-related knowledge, therefore, is a potential threat to the control that an institution has over its technology.

While the threat of hacking was potentially serious. It is certainly not a complete explanation about why CoSy users (including faculty) received little support or training in the use of the system. On one level, it might be argued that such an oversight was just bad planning. On another level, I will argue that it can also be understood in terms of the administration's view of technology and its appropriate users.

I will continue by exploring the notion of "computer literacy" and different ways of defining it, because it is a central concept at most institutions of higher education, but it often remains nebulously interpreted. The institutional definition of the concept provides insight into the bureaucratic worldview concerning technology. It provides a window into how the community college administration perceived technology, students, and the environment in which new relations of knowledge were being created—all of which contributed to whom access to information was granted.

"Computer Literacy"

> The powerful mythology of equality of opportunity and the omnipresence of the computer itself provide the most convincing justification for the necessity of CL [computer literacy]. [Noble 1984:39]

Specific technical knowledge is frequently equated with computer literacy. The "computer literacy as only narrow technical skill" approach had been adopted de facto by the community college district in which I worked. This had serious implications for the uses of technology throughout the community college system.

The district, for example, required all students pursuing a two-year degree to take one credit hour of computer course work. This suggests that the community college district officially perceived computer knowledge as important. Yet the district declined to define "computer literacy." This lack of a formal policy statement suggests that the curriculum development committee did not view "computer literacy" as a problematic concept, but rather one that was commonly held. The range of courses offered to meet the "computer requirement" further reinforced the notion of the computer as a "tool" to be mastered. Students could meet the requirement by taking a one-credit-hour course on anything from word processing to data bases. Courses were typically software specific and provided students with varying degrees of technical mastery of a software application. This created a subject who was apparently, according to the community college district, deemed computer literate (or on the appropriate path to literacy.) The result of this policy was a student with little more than a passing acquaintance with a specific application, but none of the *guiding logic* (i.e., how to critically think about a given field of knowledge and the relations it generates).

At the same time that the community college district eschewed a definition of "computer literacy," my particular college invoked a practical definition of the term. The course catalog listed knowledge of computer usage and word processing as a prerequisite for enrollment in CoSy-based courses. Lack of a formal definition left the concept open to its loose, unstated everyday usage. While the college ended up with a relatively useless screening mechanism, the attempt suggests that the CoSy administration had some strong presuppositions associated with the concept. The common idea, held by students and the community college administration alike, was that computer literacy is a narrow technical mastery of a specific commercial product, rather than a broad, interactive schema for behavior and thought.

The technical approach produces myopically formatted users who are unable to conceptualize their relationship to applications outside of their isolated area of knowledge. These individuals will remain narrowly specialized, at best, and will be less likely to challenge and redefine the social field defined by the technology and the bureaucracy behind it. The technical approach is unlikely, in other words, to efface the profile of status quo power relations defined by the computer format.

The fact that the community college defined "computer literacy" as a prerequisite for CoSy courses suggests that they were, indeed, searching for a student population embodying a certain level of knowledge. This stood in marked contrast to their emphasis on formal equality in the distribution of educational resources. In discussions by the administration concerning how to further develop the CoSy program, conversation focused on marketing to a computer-sophisticated population, especially people with telecommunications skills. In other words, the approach taken by the bureaucracy implicitly favored students who were already well prepared to use the CoSy system, in contrast with people who would perhaps receive the most benefit. I would argue that we were targeting the right student population (i.e., those who could not access courses through traditional methods, but who may have been novice computer users), but that we had an ineffective approach to the CoSy system. What is called for is a broader literacy of technology and a new way of teaching computer literacy, as opposed to an rote technical performance.

Administration Goals and Outcomes

Computer conferencing, as well as alternative educational delivery systems, were developed for two interdependent reasons: (1) these systems could reach students who could not attend courses taught in the traditional classroom setting, and (2) these nontraditional students were an untapped market. The latter consideration drove the former one. Failure to recruit and retain sufficient enrollment placed an alternative delivery program in jeopardy of cancellation.

Drop-out Rates

From Fall 1988 to Spring 1991, 49.3 percent of the students enrolled in computer-based courses completed them,[9] or 177 out of 359 students. Collegewide student retention had been much higher, averaging 84.8 percent from Spring 1986 to Spring 1990. Students drop courses for many different reasons, but the tenor of this situation was such that many of these students dropped out because of the user-unfriendly nature of the computer system and not because of course content or extraneous outside pressures (e.g., changing work schedule, family crisis). As liaison for the program, I talked with many of these students before they dropped their course(s) on CoSy. For example, one anthropology student had so much trouble "connecting" with CoSy that she dropped the course, only to pick it up again the following semester when she had more time to learn the quirks of the computer system.

Given the poor record of course completion with CoSy, it is prudent to ask why steps were not taken to intervene? CoSy courses were accompanied by a packet of materials that provided tremendous amounts of course-related information and the college's policies, but nothing more than a three-page guide on CoSy usage. Why the lack of information? I suggest that a good part of the answer rests with the presuppositions of the community college administration concerning technology and the bureaucratic logic of the college.

Labeling Students as Failures

Administrators have tended to perceive both student and computer as "logic machines" (Dreyfus and Dreyfus 1986). There is a language that constitutes this perception. Students, for example, "interface" with the computer. All the student needs to do is "master" the CoSy software to be successful, give it the appropriate "input," and get the appropriate "output." We have seen how this was inscribed in the community college's technical mastery approach to computer literacy. We have further seen through the high student drop-out rate that they were not mastering CoSy as desired by the administration.

Those students who did not finish a course were labelled as "failures." Our inadequately prepared student population had "failed" rather than the community college who provided them with their "chance." The current population could not "master" CoSy. Within this "survival of the fittest" logic, they had not survived.

"Students as failures" served as an alibi to shift their marketing target onto a computer sophisticated population, one that was "better suited" to use CoSy (e.g., not necessarily those who no other means to access classes). How did we reach them? Our strategy was to contact electronic bulletin boards, computer-users groups, and the like. One administrator noted that "techies" would take CoSy classes just because they like to work with the technology.[10] They were perceived to be the perfect pretrained population—mostly to be white, male, middle-class or upper-middle-class individuals.

How could this happen within a community college that focuses most of its energy on the nontraditional student who has no other means to get an education? At least part of the answer lies in the intersection of the language used to frame computers (e.g., the computer as a "tool") and the bureaucratic logic of the community college. Approaching the computer as a tool suggests that the community college administrators had an overdetermined view of the machine; it was ready-made, and all the college had to do was to plug in the courses and the bodies.

The logic of the institution is also relevant. The community college was a place where there were far too few people doing far too many things. There was little time to give each project the kind of attention and support that it deserved. During the 1989–90 academic year, the Instructional Technology Division's budget increased 17 percent (not accounting for inflation), the number of staff in the division decreased, while the number of students increased by 33 percent. The result was to layer more work on each of the remaining staff members. As the dean said, "We are one of the most cost efficient divisions within the college. We are so efficient, it's almost scary." "Efficiency" was a business criterion valued by the upper administration. The emphasis on efficiency carried with it "survival of the fittest" overtones because cost and enrollment are linked. Those who survived as students did so under a "regime of efficiency" that provided minimal funding and support to retain marginalized students. Efficiency and enrollment skewed services toward the already well-trained and prepared student.

"Efficiency" also meant high enrollments that translated into state funding for the college. Programs that did not deliver high enrollments were candidates for cancellation. Because time or resources were not allocated for evaluation and adjustment program delivery, it became necessary to target a population that would not drop out. Increased enrollment (within the efficiency algorithm) meant legitimacy regardless of which social sectors that enrollment came from.[11] The CoSy administration reshuffled its interests and goals in order to bring stability to the program and satisfy the upper administration's short-term goals (cf. Latour 1987:113).

In sum, the administration attempted to bring success to CoSy through a strategy of displacing the goal of reaching students who *need* to access courses through nontraditional media toward a goal basic to the college—increased enrollment, or the "bottom line." The administration translated their shifting goals in terms of student "failure"—thus assuring that the community college administration was not implicated in the drop-out problem. It was the student who failed, not the college.

Conclusions

Computer conferencing can provide students with access to otherwise inaccessible educational resources. However, helping the nontraditional student gain access to an education through the use of computers had little to do with how the technology was distributed. Students at the community college had ample access to hardware in state-of-the-art computer labs if they lacked their own machine. Two major factors determined who got access to the educational resources provided on CoSy: (1) the format of the technology, and (2) the institutional logics and the practices that stemmed from those logics. Technology both shapes and is shaped by the institution that uses it (Barley 1986).

This article has been critical of the deployment of technology in the educational process but should be seen as a cautionary statement and not as a global indictment. My goal, rather, has been to alter our assumptions about technology in a way that will contribute to more successful applications. To realize this, it was necessary to unpack some of the important management logics (i.e., institutional worldview) that have shaped the deployment of CoSy at the community college. These include an emphasis on short-term gains versus long-term planning, enrollment and efficiency, technology as a tool, and computer literacy as a form of narrow technical competence. In sum, technological formats had an interpenetrating relationship with the logic and practices of the institution. What bound them together was the process of decision making by the community college and its effects. Taken together, these effects, in conjunction with the format of the technology, resulted in the exclusion of novice computer-users from accessing CoSy-based courses. These institutional effects limited the probabilities for the success of CoSy or any other conferencing package that might be employed. Conceptual separation of the technology from the institution could lead to premature and incorrect recommendations about the CoSy system. Such a position would suggest, for example, that a change of software would solve the problems experienced by computer-conference students.

In the Fall of 1991, a change in software did occur. CoSy was replaced with a locally developed computer conferencing system known as the Electronic Forum (EF). The deployment of EF was backed by two years of design and development, as well as extensive field-testing and feedback within the community college district. Unlike CoSy, students and faculty found EF simple to learn and easy to use. If our hypothesis concerning a change in software were true, then retention should have risen in the Fall of 1991. It, in fact, descended below the general average of 50 percent to 44.6 percent. How is the on-going failure of this computer conferencing system, in the light of better software, to be explained?

The technology changed, but the practices and logics in which it was embedded remained the same. CoSy was a sign for institutional logics and a site where those logics were enacted. EF was drawn into a similar logic and was subject to a similar set of practices.

An important theme in this article has been the overall lack of commitment and resources for the computer conferencing program. A site license for CoSy was inexpensive; and when it came to replacing it, EF was chosen because it was free. (In short, a very narrow cost-benefit analysis was used.) As a result, it was an accident that the college now uses a well-designed conferencing package. At the same time EF came on-line, other processes were in motion that undermined the potential success of the new system. Students were provided with a basic telecommunications software package in order to standardize the dial-up process for home-based students and make it easier to work with students experiencing problems. This software was chosen not for its intrinsic quality and user-friendliness, but because the college already owned a site license for the software—again signifying a narrow cost-benefit analysis. The software was unstable. It worked on some machines but not others, and often locked up while uploading files. It often refused to hang up phone lines. Finally, scripted files designed to dial and connect students directly into EF worked erratically.

Phone-line problems further compounded student frustration with the new EF system. Students attempting to dial up and connect with EF would either get no connection or a busy signal. When they did connect, the transmissions were so "dirty"[12] that students could not work in their computer conference and were frequently disconnected. While a problem with the phone lines used for administrative mainframe access was fixed in a matter of days, problems with the student phone line continued for a month.

Students using one of the colleges microcomputer labs faced other problems. They had access to microcomputers that connected directly to the mainframe through a network, making for a consistently reliable connection to EF. Nevertheless, the retention of lab-based students was lower than for their home-based counterparts: 54.8 percent of those working at home completed their courses as opposed to 46.2 percent of those working at computer labs. The college explains this in terms of less sophisticated computer-users working at lab sites who have more trouble with the technology. Ironically, it could also be hypothesized that lab-based students should exhibit higher retention because they have access to technical support staff. As liaison, I ended up working with many of these lab-site students. Their stories were consistent. Lab personnel did not understand how the computer conferencing system worked and could not answer basic questions. Attempts to schedule staff-training sessions at the labs were refused. Ultimately, I worked with several frustrated lab personnel on an individual basis, but support for computer conferencing was not institutionalized at the lab sites through regular training sessions.

The transition from one technology (CoSy) to another (EF) provides an interesting contrast to the work by Barley (1986) who examined how the same technology (CT scanners for medical imaging) can have strikingly different organizational effects in two different institutions. Technology and its use becomes defined and constituted by the context in which it is located. The majority of this article focuses on one technology, its organizational effects, and its ultimate failure. The recent introduction of EF into the community college points toward the conclusion that different technologies can have similar outcomes when deployed within the same organization.

The usability and success of CoSy, EF, or other computer conferencing systems, consequently, is dependent upon the organization that adopts and deploys them. The introduction of computer conferencing into an organizational setting with a history of divisive political and power relations has, in the case of CoSy and EF, contributed to their failure. As Dubinskas (this volume:1) observes,

> The inability to predict social consequences from the technology alone necessitates a detailed social study of implementation contexts to help determine the dynamics of the process. At the same time, the character of a technical system does have an *influence* in shaping the possibilities of its use.

The interpenetration of technology and institution results in the creation of new social constructs. In the case of the community college, new constructions involving people, machines, and practices generally conformed to the institution's dominant logic and procedures. Barley (1986:80–81) notes, however, that "slippage" occurs that alters the existing structure and relations of power. In Barley's case, he is examining how technology mediates face-to-face relations between technicians and physicians. Technicians running the CT scanners had knowledge that radiologists and physicians did not, thus skewing traditional power relations in the hospitals he studied. In the computer conferencing system described in this article, people were structurally distanced. Students were not in the same circuits of power as the hospital employees of Barley's study. They did not control important knowledge or resources. In the community college context, those who succeeded had the requisite skills and flexibility to work with the equipment. Others, through persistence, managed to complete their courses. CoSy and EF are systems characterized by low intentionality— that is, no one set out to create a poorly designed system. Yet, events emerged that conformed to a general set of institutional values and practices that undermined the success of the system. In a system characterized by so much contingency, students succeed by accident. As Jackall (1988:202) notes, rational systems become subject to private and organizational agendas, intraorganizational squabbles, and random events that often produce the opposite of what was intended.

Notes

Acknowledgments. I owe debts of gratitude to a number of individuals who have contributed to the final version of this article. Frank Dubinskas, the co-editor of this volume, has provided continual support since our initial discussions for a special session at the 1990 AAAs. I am also particularly indebted to Dion Dennis, who provided insights and feedback throughout the many drafts of this article. Thanks also go to Kay Sands, David Altheide, and the NAPA reviewers for critical readings of earlier versions of this article.

1. Version 2.04.

2. Even though the community college offers classroom-based courses at nontraditional times and locations, access to these classes requires that a student be able to be in the classroom at predetermined times. Students with irregular or unpredictable schedules (e.g., swing shifts, heavy travel) and other constraints (e.g., child care, physical disabilities) often cannot conform their schedules to the demands of a classroom-based course.

3. Alienation is a complex and post hoc label that encompasses a broad range of student experience. It includes the student's sense of being conceptually "lost" within an unknown piece of computer software. It is also the sense of being physically dislocated. These students have been culturally imprinted with the notion of the classroom as a physical location filled with known individuals. Placing them in an electronically mediated educational environment leads to feelings of frustration and confusion. I would also argue that a sense of alienation results from the technology being treated as though it were a "glass tool" when it is, in fact, so obviously a part of the class. Instructors on the CoSy system would do better to incorporate the computer network into their curriculum rather than deny its presence, or treat the issues of technology and course content as two distinct and independent domains. Indirectly, then, teaching techniques also contribute to the student's sense of alienation. Finally, while all of the CoSy-based instructors would like to think that their course material was nothing short of riveting, lack of the human experience of the classroom and the performance involved in teaching (at least if it is done well) may leave students bored. These various components are undergirded by a common theme of *connection*. What alienates students is their inability to connect the cold bureaucratic communication format with a more private, intimate, and emotional communication format.

4. "Advantage" refers to a student's skill and general comfort using advanced technology, such as the computer. These characteristics tend to follow gender and class contours.

5. This part of the classroom discourse might best be thought of as a form of "intimate knowledge" that makes a class dynamic. In a computer-based course, the intimate knowledge portion of the discourse is missing. It impoverishes the experience.

6. For those unfamiliar with the technological terminology, this is the use of a modem and communications software to connect a home computer to a distant computer via existing phone lines.

7. This distinction between "hot" and "cold" events merits some further discussion. Bureaucracies produce and circulate information in conventionalized formats. In this formal public realm of information, format and structure dominate bureaucratic discourse. This is a "cold" domain of communication. This can be contrasted with a private, informal realm of communication, which is highly contextualized, emotional, and meaningful for those interacting. This is the "hot" domain of communication. CoSy provided an arena in which communications were highly structured (and closed) in their written form. There was none of the spontaneity and emotion of spoken discourse. A live lecture for example, although structured along certain conventional lines, has an energy and emotion that the same lecture in printed form cannot reproduce. Written interactions (especially those that take place between people who do not know one another, as is the case with CoSy students) generally lack the tonal qualities of speech, as well as other nondiscursive elements of live interaction.

8. In the electronic classroom, students were expected to make some contributions toward the discussion of weekly materials. Their commentary was not deleted once it was read; rather it stayed on the system until the end of the semester. In a traditional classroom, students may also make comments in a class discussion, but their contributions are far more transitory and far less subject to on-going examination than written comments. Consequently, students were being asked to go much more "on the record" in the electronic classroom than in its traditional counterpart.

9. This category includes only those students who remained in a course after the drop/add registration period during the first week of school.

10. This does open up the possibility to expose people to new content areas toward which they might not normally gravitate.

11. In discussions with the associate dean, it became clear that the upper administration had put no direct pressure on her to "make CoSy work." On the other hand, they had not

backed the CoSy program with the institutional support and funding it required. The dean's response was to "get on" with the program given the fact that the necessary support for it was not forthcoming. In this case, "getting on with it" meant finding a definable and reachable population that would be receptive to CoSy. Recruiting the technically proficient electronic bulletin board user was "efficient."

12. A dirty phone line is characterized by static that is often imperceptible when listening on a standard telephone receiver. A computer will translate this static into computer characters.

References Cited

Altheide, David L.
    1985a   Media Power. Beverly Hills, CA: Sage Publications.
    1985b   Keyboarding as a Social Form. Computers and the Social Sciences 1(2):97–106.
    1989   Formats of Control and the Self. Studies in Symbolic Interaction 10:187–197.
Barley, Stephen R.
    1986   Technology as an Occasion for Structuring: Evidence from Observations of CT
        Scanners and the Social Order of Radiology Departments. Administrative Science
        Quarterly 31(1):61–103.
Bowers, C.A.
    1988   The Cultural Dimensions of Educational Computing: Understanding the Non-Neu-
        trality of Technology. New York: Teachers College Press.
Dreyfus, Hubert L., and Stuart E. Dreyfus
    1986   Mind over Machine: The Power of Human Intuition and Expertise in the Era of the
        Computer. New York: Free Press.
Jackall, Robert
    1988   Moral Mazes: The World of Corporate Managers. New York: Oxford University Press.
Kidder, Tracy
    1981   The Soul of a New Machine. London: Allen Lane.
Latour, Bruno
    1987   Science in Action. Milton Keynes, England: Open University Press.
Noble, Douglas
    1984   The Underside of Computer Literacy. Raritan 3(4):37–64.
Poster, Mark
    1990   Words without Things: The Mode of Information. October 53:63–77.
Walters, Andrew J.
    1988   First Time Computer Users. Ph.D. dissertation, Department of Education, Arizona
        State University.
Wolf, Eric
    1990   Distinguished Lecture: Facing Power—Old Insights, New Questions. American
        Anthropologist 92(3):586–596.

When Freedom of Choice Fails: Ideology and Action in a Secondary School Hypermedia Project

Gail Bader and James M. Nyce

Introduction

Recently, hypermedia has become important in the educational community and for the developers and vendors that work with it. This is because hypermedia has been both developed and promoted as a resource that enables (or at least creates the circumstances for) "self-learning"—learning in which students have both control and choice over how and what they learn.

Here, we will argue that this ideological presentation of hypermedia is both powerful and attractive because, in part, it clearly articulates an important cultural understanding of what education in the United States should be—the individual learning to make choices. As a resource and a technology, hypermedia then promises (or at least may allow this) to transform what have been traditional student-teacher relationships. Hypermedia offers, its advocates argue, a chance or opportunity to move from models of education that are teacher, not learner, centered. This ideology claims to invert, if not to transform, central informing relationships and hierarchies in American education.

We will suggest that as powerful a cultural appeal as this inversion is, this appeal is doubly compelling because it offers the possibility to hide or to conceal hierarchy. While this is not the same thing as dissolving hierarchy, it can be argued that in American terms this is the next best thing (Varenne 1974). We will argue that, in fact, regardless of the rhetoric and arguments made for hypermedia as an educational resource, it will neither subvert nor transform hierarchy in the education system nor the relations of power engendered by and embedded in that hierarchy. It merely obscures it. In practice, however, these two opposing cultural principles, hierarchy and choice, are alternatives with which teachers must then deal.

The Selling of Hypermedia

Hypermedia programs are software applications that enable users to create machine-based ties or links between disparate materials (this may include video, graphic, textual, and audio materials) in the form of "files." These links allow the user to travel between files vis-à-vis these links. More than simply allowing users to connect different computer files, hypermedia touts that "connecting" these files has the potential to reconfigure knowledge in a new way. As Virginia Doland has pointed out: "A 'link' posits a relationship between two nodes containing information, and thus creates a new intellectual entity, an assertion about reality which, accepted or not, does not leave the reader unaffected. Simply to create a linkage is to create a unit of meaning" (1989:10). Thus, the "promise" of hypermedia applications is the potential to reconfigure knowledge in personally meaningful and satisfying ways through this linking.

As compelling as the promise of restructuring knowledge in personally meaningful ways is, we believe an essential element of this reconfiguration of knowledge is that it ideologically supports individual choice. While much has been written about how important individualism is in America, little work has been done on what this means in terms of theories of action (but see Varenne 1977:53). In his study of American life, Varenne suggests "that the child is already a fully formed person who can make choices based on rational and intellectual premises, a person who possesses the inner wisdom to teach himself" (1977:43). This notion of the person, he believes, informs how Americans believe children should be educated. "Real teaching" (as opposed to education) for Americans takes place when students (or others) are led to discover the truth for themselves (Varenne 1977:44). It is this particular notion of individualism and education that makes the ideological promises made for hypermedia so powerful and persuasive.

Those who claim that hypermedia programs are "different" from other educational software programs do not do so simply because these programs allow users to represent, select, and link materials. Rather, the technical ability to do these things is framed as representations of individual action and freedom of choice. This freedom of choice is embedded in the technology of hypermedia via the selection of material (the user chooses among the available files), links among material (the user chooses which links to follow), and the interpretation one can draw from the files and links he or she has chosen.

To make our point clearer, we must look at how other educational software programs are framed by those who argue for hypermedia. They believe that drill and practice or tutorial programs are prestructured programs that allow no choice on the part of the student. In the world of educational software, then, programs like "drill and practice" set the backdrop for the supposed virtues of hypermedia—that is, it is "structureless" and becomes structured through the choice (therefore action) of individual users. Again, we return to the promise of hypermedia—that with it, individuals can represent their own personally defined version of what knowledge is.

As educational software, hypermedia, with its appeal to individual choice, may be particularly attractive because of current definitions of what constitutes an "authentic education" in America, which Varenne ties directly to choice and action (1977:42–45). One of the important ideas about education today is that education should be neither mere rote learning nor memorization. Rather, it is something that actively engages the individual and involves the learner in his or her own experience of education. In part, today's concern with "engagement" and active participation is a result of the American commitment to the idea that individual action validates and reinforces authentic education. In American terms, it also reflects our concern with individualism. For education to succeed, as Greenhouse has pointed out, individuals must strive to perfect their knowledge, and they can only do this themselves (1985:262).

This concern with the learner's active engagement in the educational process, whether that process refers to classroom strategies or educational software, presents itself in contrast to how education is conventionally structured and understood in schools. This contrast and the concern with the learner's active self-engagement was a central feature of the hypermedia development project we studied in two secondary schools.

The ACCESS Project

In 1989, we began the second year of research in a three-year project using Apple Computer's HyperCard to develop an interdisciplinary corpus (ACCESS), linking American history and literature. The goals of the project were to make the relationships between concepts in history and literature explicit; to permit students to acquire new concepts and information; and to help students explore and construct relationships between concepts. This year, we worked with a history and literature teacher at two high schools. We looked at how these four teachers went about creating materials for a hypermedia corpus and how they used it in the classroom. We also interviewed and observed their students (typically juniors), looking at how they understood and used the ACCESS materials.

In this project, the ACCESS corpus was framed as striking a powerful blow for individual choice and authentic learning. The teachers not only hoped to encourage a new way of thinking about American history and literature, but they also hoped to encourage a certain kind of learning experience—that is, one that engaged the individual and focused on the configuration of knowledge in individually meaningful ways.

As powerful as this technology promised to be in terms of encouraging student choice and allowing them to reconfigure knowledge in personally meaningful ways, this promise was significantly constrained in practice. Indeed, we believe that the whole question of "choice" was contested and redefined as the technology was used in the classroom. In the following section of this article, we want to look at one incident that raised the question of the role "choice" plays in classroom practice.

ACCESS, Choice, and Knowledge in the Classroom

One of the teachers involved in the project was particularly interested in encouraging her students to develop and follow their own insights when working with the hypermedia materials rather than simply recreating her own point of view. We call here the kind of learning experience she talked about "self-discovery." In developing her strategy for the hypermedia materials, she assigned these materials to students after they had read particular literary works but before she lectured on corresponding literary concepts and traditions. Here student "choice" was linked to the notion of "exploration" among the many potentially relevant materials and, as in her assignment on the Romantic Period, exploration meant encouraging the students to "Feel free to pick and choose . . . [among the many such available materials on the system] and follow LINKS that interest you" (English III H handout). She hoped that the hypermedia materials, assigned this way, would allow the students to develop and support their own definitions of Romanticism. In important ways, self-discovery meant that a learning experience showed student engagement that was not mediated by the teacher.

Generally, she was pleased with the extent to which her students connected materials in original ways. In short, because students were offering "unique" insights—that is, they were making connections she had not herself pointed out to them—she felt some self-discovery was occurring. However, a problem arose. What happened when these "unique insights" were wrong or only partially correct? For example, when she asked students to explore the ACCESS materials marked "Romanticism," she also asked for written responses to one of two questions, one of which was "Is Washington Irving a 'Romantic'?" After reading student responses to this question, she discovered that the students had not recognized that Irving's work combined elements of both Romanticism and satire. In short, her students had discovered only "half" of the correct answer.

This instructor now faced a problem. On the one hand, she had hoped to create a learning situation in which students experienced "learning" as an exploration of their own ideas and interests. She felt she had some success when the students made connections she had not seen or had not talked to the students about. Here, however, she was confronted with the absence of a "connection" (i.e., Washington Irving's work has both Romantic and satirical overtones) that was important enough for her to cover the issue again in class, to change her grading criteria for the assignment, and to rethink her strategy for developing questions. This raises yet another issue. Why were the students' "choice(s)" considered "wrong" or only partially correct?

The teacher, although truly concerned with encouraging students to experience learning as the engagement of self, discovered, somewhat to her dismay, that she did indeed have an agenda of her own that sometimes took precedence over student self-engagement. Her personal commitment to teaching and to "established knowledge" would not allow her to let students believe that they were (completely) right in an answer they could get "wrong" on various college entrance exams or in other classrooms. Although she wanted the students to experience the hypermedia material themselves, she could not let this stand in the way of students learning and having "essential" knowledge. This assignment then created a dilemma for the teacher and the students in which "self-discovery" and "correctness" collided in the classroom. However, when push came to shove, self-discovery gave way to essential knowledge.

One of the most interesting aspects of this situation however was the teacher's assessment of why it had happened. Rather than questioning the role "choice" had in learning, she attributed the "incorrectness" to the way she had phrased the assignment question. Instead of asking if Washington Irving was a Romantic, she felt she should have asked the students to show how Irving's work was both satirical and influenced by Romanticism. Clearly the two different ways of asking the question reflect a difference in the extent to which the teacher structures the students' view of information, and she was aware of this. In this situation, the teacher had begun "repair work." She set about restructuring the students' own, independent understanding of the material and so redefined the notion of choice in the classroom.

Context, Practice, and the Notion of Self-Discovery

As we have pointed out, what seems to drive American education are certain deeply embedded notions about the self as an active agent and consequently that the best, most valid, kinds of learning occur through an active engagement of self. What we have here is an occasion to consider how these notions are articulated in practice and the role that context and certain kinds of structure have in the defining of these concepts. In short, self-engagement neither emerges from nor operates in a vacuum. What we have tried to point to here are some of the conditions that determine how these concepts about self and education are defined and worked out on the ground.

What we did not find in this project is some kind of fundamental discontinuity between the principles and assumptions embedded in hypermedia and espoused by hypermedia developers and those that underlie the rest of the work these designers and teachers do. The same assumptions and principles in fact run through, underlie, and to a large extent, determine all of the work these teachers do.

The difference is that certain constraints and structures, particularly those that are classroom and course specific, intervene and mediate these concerns with self, action, and self-discovery. To be more precise, these structures do not just constrain classroom practice. Teachers see them, if not as immutable, as largely outside of their control. It is this definition of the situation—that elements of the institution, curricula, and classroom practice are set in place and have to be taken for granted—that restrains and in the end works against these concerns with self, self-discovery, and active learning.

Conclusion

Hypermedia applications imply that hierarchy and authority have no place in the educational experience. Moreover, they imply that there are no intervening features in the construction of knowledge other than the features the individual himself considers important. Underlying these claims is another: the use of hypermedia resources in education will change, if not, transform established social relations in the classroom in that it will "empower" students and learners. In fact, what we found was something quite different. Hypermedia, both as a corpus and in use, reflected and became redefined in terms of established educational practice.

More than any other educational technology today, this is the promise that hypermedia holds out. In other words, it offers an arena that will promote and provide for self-discovery. What is implicitly denied here is hierarchy (who knows best) and structure (what students should know), and it fails for just these same reasons. What is not taken into account is that classroom practice bounds what constitutes "legitimate learning." Equally important—for this says something about American notions of individualism—the project failures of this kind were not recognized or even talked about in terms of hierarchy or structure. Rather, the teachers turned from hierarchy and structure and sought explanations for the project's failures and shortcomings in themselves and their students.

Hypermedia sells on the basis of a learning experience linked to self-discovery, one that promises choice and so, in American terms, yields authentic learning. In this project, teachers did not reject the value and worth of self-discovery. Instead, they redefined it in particular ways, and this had certain consequences for the project and the corpus itself. For example, there was a shift from a view of a hypermedia resource as an individual student resource (capable of promoting an individual authentic learning experience by itself) to a resource for the course and class discussion. To put it another way, because of the demands of classroom practice, with ACCESS, freedom of choice still figures but this body of knowledge, this corpus, is no longer entirely self-directed or self-determined.

Teachers originally hoped ACCESS would be a resource for genuine self-discovery. They had taken for granted the idea that "authentic self-discovery" alone could and would support the aims and goals of a course. However, when students "discovered" the wrong answer, teachers began to frame, stage, and manage "self-discovery" via assignments, classroom discussions, and directions for use. In short, the acquisition, representation, and valuation of knowledge that came to constitute self-discovery in hypermedia was determined and mediated by the kinds of knowledge seen to be relevant to the aims and needs of a particular course, class, and curriculum. In the end, this hypermedia program became valued for how it could be used to further the instructor's aims for a particular course rather than for what it offered individual students.

Even though we have reservations about how hypermedia systems are being used today, we still see a great potential for their use by teachers to teach in all sorts of courses a number of skills that we associate loosely with "critical" or "analytic" thinking. Our primary concern is that teaching of this kind should be incorporated into the use of the systems in teaching contexts and that the teachers, when using hypermedia in the classroom, make their reasoning and arguments explicit. Only when this occurs, can students really work through for themselves how these associations and links are constructed, how these arguments are made, and how to evaluate the evidence used.

Our point is that hypermedia systems are teachers' tools. In short, they must come to be seen as tools, much like the blackboards that teachers use everywhere. Nevertheless, what has to be kept in mind is that typically when teachers use blackboards, they recognize that what they write on the board is not the same as (is not a substitute for) their discussion of whatever those points are. Teachers need to focus (as do vendors and researchers) on hypermedia systems as tools to support their teaching (e.g., ones that make their intellectual positions clearer, more precise, and more accessible to students) rather than assuming that this will occur "naturally" in the process of contributing to or developing a hypermedia data base.

Acknowledgments. This work was supported in part by the James S. McDonnell Foundation, the U.S. Department of Education Center for Technology in Education and Apple Computer, Inc. We wish to thank Donna Muncey and Bill Graves for their comments on drafts of this article.

References Cited

Doland, Virginia D.
    1989   Hypermedia as an Interpretive Act. Hypermedia 1(1):6–19.
Greenhouse, Carol J.
    1985   Anthropology at Home: Whose Home? Human Organization 44(3):261–264.
Varenne, Herve
    1974   From Grading and Freedom of Choice to Ranking and Segregation in an American High School. Anthropology & Education Quarterly 5:9–15.
    1977   Americans Together: Structured Diversity in a Midwestern Town. New York: Teachers College Press.

Commentaries

Romancing the User: Hi-Tech Teaching in Anthropology and Industry

Anna Hargreaves

Upon hearing the six papers at the American Anthropological Association session on "Closing the Educational Gap: Alternate and Hi-tech Teaching Methodologies and Their Implications," I was immediately struck by the parallels that exist between the academic environment and the industry environment in which I work. A major focus of my job at Lotus was designing end-user training for software applications. Until recently, the courses were largely paper based. Lotus is now moving rapidly into the area of on-line, interactive training and hopes to gradually offer multimedia courses with sound and animation, in addition to the text and on-screen graphics that it currently uses to deliver training for its software products. When you adopt hi-tech methodologies, whether it's to teach people to become anthropologists or software gurus, many of the same hurdles have to be crossed. Hi-tech delivery media are still in their infancy, and we are just beginning to recognize what makes them seductive, but also problematic.

I would like to address some of the ideas in this volume that deal with the advantages and limitations of hi-tech as a delivery medium for teaching anthropology and to add my own commercial developer's perspective to the discussion. Everyone seems to agree that, like it or not, hi-tech instruction is here to stay, but few would describe it as a resounding success.

Looking first at the positive side, the one clear advantage of the hi-tech medium is that it provides a convenient way for students who might otherwise be excluded because of other commitments, time conflicts, or physical distance, to take courses. Students can learn in their own time, at their own pace, and from their own armchairs or villages. More students can be reached this way, and for cost-conscious administrators, increasing enrollment with a smaller faculty is an excellent proposition. In the commercial software development world, clients are also requiring us to provide alternatives to the expensive and impractical classroom training sessions, as well as snazzier supplements to our traditional printed documentation. In release 2.3 of the Lotus 1-2-3 product, we provided an interactive, on-line tutorial that puts students through the paces of using a spreadsheet. In future releases of its products on several different platforms, Lotus will be offering more elaborate electronic tutorials and user-support systems, such as extensive, multilevel help, on-line manuals, and sound-enhanced guided tours. Another advantage that comes through in several articles is that done right, hi-tech courses can be enjoyable and can actively engage students in the learning process. Good courseware can do more than teach a set of facts about the way we currently do things; if the technology is sophisticated enough, it can lead students into unchartered territory and encourage them to create connections of their own. Bader and Nyce give a good example in which students create their own links between literary and historical events of a given era through the use of a hypermedia program. The parallel at Lotus is that clients do not want canned, generic solutions to business problems—more and more they are looking for ways of applying their knowledge of a new software system to their own applications. They are seeking not only to acquire technical skills, but to expand this knowledge and to deploy it strategically. In addition to "how to" modules, we are providing more and more application-oriented scenarios and templates that suggest creative ways of using the software skills on the job, whatever that job might be.

Courses that encourage student interaction and self-discovery are the essence of what hi-tech can add to traditional learning methods. Courses that most closely mirror the real world, like the Fugawiland archaeology simulation, make full use of the hi-tech medium. They present different approaches and choices that prepare the students with intellectual tools for life beyond the learning experience.

The fact that the students can reconfigure knowledge in personally meaningful ways is a powerful concept that may not always have the expected results, as Bader and Nyce propose. If individual choice is carried too far, it can lead to difficulties and misinterpretations. This is certainly true, too, of the more defined domain of software training. While our goal is to give users as much control as possible over their learning environment, giving them total freedom would lead to confusion and often to incorrect results. The trick then is to gauge how much freedom to allow, and this is not an easy call from either the instructor's or course designer's point of view. The degree of flexibility you can give to the student also depends on the software's capabilities. While hypertext systems are able to do this, many of the lower-end products provide a strictly linear approach that cannot be easily customized and is therefore inflexible for both the student and the instructor. The other side of the coin, as Dubinskas suggests in his introductory remarks, is that flexibility is not a virtue in itself; students have to understand how to make use of it in order to benefit from it.

One of the pros of hi-tech systems discussed by Hamill and Marchant is that attractive graphics (even if they cannot be easily printed!) and animation can greatly enhance the learning experience, and there is no reason why learning should not be fun. In industry, we put a great deal of emphasis on the aesthetic and graphical appearance of our training programs, because users have come to expect this with the evermore present Graphical User Interface technologies. A bland presentation no longer holds up against the competition. Do these advantages outweigh the problems we have today in designing and teaching hi-tech courses? I think we have a way to go before we can say with a clear conscience that the courses offer real added value to students over conventional teaching methods. But eventually, with advances in technology and a better understanding of the "end user," the new electronic delivery systems should have a more secure place in the teaching curriculum. The articles in the volume provide real-life examples of this theory; and in industry, we could learn from this style of research. As Brenda Laurel says in her volume on human-computer interface design: "observations from one user/context intersection . . . can inform the design of interfaces for other populations and tasks. Like cultural anthropologists, we will often find native informants to be indispensable in guiding our explorations" (Laurel 1990:93).

The problems are complex and challenging, particularly when you are trying to teach subjects like anthropology in which social implications and value systems are an integral part of the process. The awareness of social issues may be less pressing in the case of software training, but it is becoming increasingly important in the face of growing international markets and diversity issues in the workplace, both important business themes of the 1990s. Industry has a great deal to learn from the attention anthropologists pay to cultural issues in their teaching methods.

One of the main problems expressed, in particular by McDonald, and to some extent by Hamill and Marchant, are the technical shortcomings of many of the delivery systems currently available at an affordable price. The articles describe some of the programs as clumsy, unintuitive, buggy, and designed by programmers rather than instructional specialists. As long as this is the case, student attrition rates will most definitely continue to rise. There are still too many situations where the technology is cumbersome and intrusive. Until the delivery medium becomes transparent, students will remain dissatisfied. In software training, the medium and message may be closer, but the students' expectations are extremely high. Quality and user-friendliness are characteristics that software trainees have come to expect, and we still fall short in meeting these criteria. One of the obvious solutions that is too often neglected is extensive user testing. As Vertelney, Arent, and Lieberman (1990:53) conclude, "The role of well-planned and well-executed user testing throughout the design process cannot be overemphasized."

Another problem on the technical front that still looms large is the fact that students have a hard time navigating around the course materials and finding what is really relevant. Mason Weiss, Metzger, and McDonald and Segal both refer to this phenomenon. As one of Segal's telecourse students says: "it's a real crapshoot out there figuring out what's important and what's not." This is a common problem we have to deal with in designing software training. The more dense the material you are presenting, the more important it is to provide clear navigational cues, maps of where you have been and where you are going, and the ability to ask for context-sensitive help when needed. We search constantly for good metaphors that convey, via analogy, how to get from A to B, how to find your way around a maze, or how to get yourself out of trouble. As Joy Mountford (1990:25) of Apple Computer, Inc. suggests, "metaphors are powerful verbal and semantic tools for conveying both superficial and deep similarities between familiar and novel situations." She goes on to note that:

> Much of the ease of use of the Macintosh interface is attributable to the correspondence between the appearance, uses, and behaviors of such interface objects as documents and folders and the real-world counterparts; conversely, user difficulties are often attributable to differences between them. [Mountford 1990:25]

We also need to find visual ways of highlighting important information, and of prioritizing and organizing the information on-line, as Mason Weiss, Metzger, and McDonald note in their case study. However, rather than finding clever fixes for existing programs in order to provide better interfaces for students, we should concentrate on building these features into the programs themselves. Instructors and students who work around the program inadequacies deserve a prize, but we should be addressing these inadequacies up front. An inspiring resource on the art of visual presentation techniques is Edward Tufte (1990) who sees great hope in the high-resolution computer visualizations of the future as a way of making the "flatlands" (paper and computer screens) more intelligible to users.

We also need to find better ways of giving feedback to students when those SOS calls are heard. Built-in diagnostics and error-recovery mechanisms should be commonplace, to avoid the frustrations that McDonald describes. This is especially true of courses that are designed specifically for self-study—far from any kind of immediate assistance if students run into technical difficulties. More telephone help hotlines and on-line tutorials would allay the current fear of technology that mars students' experiences in distance learning programs.

The need expressed by some students for more personal contact with peers and professors is fairly widespread but seems in a sense to go with the territory. You cannot enjoy the geographic independence that the telecourses offer and at the same time greatly increase the amount of personal contact between students. This would defeat the raison d'être of many telecourses that are designed to reach a time-constrained and dispersed audience, but making it easier to interact via the technology and discuss issues on-line might help alleviate the problem.

Another salient issue that has less to do with the technology than with the design of the course, is the fact that the materials are most frequently someone else's agenda as Segal laments. If the telecourses are used for self-study, as they often are, the students get a single perspective (and maybe a misguided one). There is less opportunity to discuss other points of view, to socially construct a body of knowledge, and therefore programs run the risk of promoting "ethnicism"—the denial of local cultural specificity or difference.

Although the designer's point of view may be less controversial in the context of software courses, I think this is an area that deserves more attention. In choosing examples to present, we need to be culturally aware and sensitive. This is not only a courtesy to our international audiences who now make up over 50 percent of the market, but also a cost-saving measure in the long run when we have our products translated into other languages.

Considering the many problems outlined here, we need to be judicious in choosing how to use hi-tech in teaching. With the advent of more and more attractive media for presenting course materials, we need to continually ask ourselves, "what's the value added from the student's perspective?" If there are good reasons for using hi-tech, whether it be video, hypertext, or telecourses, we should be sure that we are delivering quality products, with adequate support and training, so that the medium does not intrude on the message. The guiding logic that McDonald refers to is a key concept. As course providers, we must find ways to instill students with the confidence and expertise to overcome their fears by offering good support systems that demystify the black box. There is a price to this that we should factor in when we budget for hi-tech courses.

Two more observations drawn from my own experience might suggest some direction on how to move forward. The first is the importance of having real commitment and resources. Whether we use off-the-shelf courseware or off-the-shelf tools, such as hypertext, to develop our courses, we should look for the ability to adapt or extend the tools. This may seem Utopian in the face of constant cutbacks, but it is our best hope of success. Nothing prepackaged will ever perfectly suit the needs of any particular group. The solution is to find something that covers at least some of your specific needs and that can be improved to meet these needs more closely. With finished courseware, you probably would not have the license to make changes, in which case it is worth considering a friendly relationship with the courseware providers. If you do have access to the technology, as in the hypertext example, then you may require some technical assistance to make the adaptations. Ideally, the tools you use should be flexible, and some level of support and technical advice should be close at hand.

My second observation is related to the emergence of groupware products. CoSy is one of these products, in that its goal, at least superficially, is aimed at facilitating group communication. I think these tools will become much more user-friendly and popular over time. Designers will develop better intuitions about the tools necessary for groups so that the tools not only tolerate group communication, but also enhance communication and change the way people work.

We will also start to see tools that were originally designed with the single user in mind but have been adapted to groups. This is already happening in the world of software applications at a very rapid pace, and the same could happen with other types of programs. Telecourses could be designed to represent multiple perspectives by creative use of existing technology. The whole movement to CSCW (Computer Supported Collaborative Work) is indicative of both academic and industrial interest in this area (Greif 1988).

Lotus is experimenting with this concept in the design of its software applications and the accompanying on-line training. The applications themselves will have better mechanisms for tracking who does what or says what in a group environment, and in the training materials, Lotus offers several layers of training, depending on the knowledge level of group members. You can choose a beginner's track, or a more advanced track, assistance via text or sound, and other such personal preferences that are the fabric of any group.

These systems are complex to build, but they come much closer to reflecting the reality we are trying to convey regardless of whether the subject matter is Anthropology 101 or 1-2-3 spreadsheets. We can take solace in the fact that experts do not regard system design as easy either. Thomas Erickson (1990:3) gives us three reasons why interface design is such a challenge. "First, quite simply, it's hard to come up with good solutions. Second, there are so many competing desiderata involved in interface problems that any solution is bound to be a compromise. . . . The third reason for the difficulty is that it's interdisciplinary and highly political." When we have jumped over some of these hurdles—and we will—we will be able to pronounce hi-tech teaching methods a real success.

References Cited

Erikson, Thomas D.
    1990 Creativity and Design. *In* The Art of Human-Computer Interface Design. Brenda Laurel, ed. Pp. 1–4. New York: Addison Wesley.
Greif, Irene, ed.
    1988 Computer Supported Collaborative Work: A Book of Readings. San Mateo, CA: Morgan Kaufman Publishers.
Laurel, Brenda
    1990 Users and Contexts. *In* The Art of Human-Computer Interface Design. Brenda Laurel, ed. Pp. 91–93. New York: Addison Wesley.
Laurel, Brenda, ed.
    1990 The Art of Human-Computer Interface Design. New York: Addison Wesley.
Mountford, Joy
    1990 Tools and Techniques for Creative Design. *In* The Art of Human-Computer Interface Design. Brenda Laurel, ed. Pp. 17–30. New York: Addison Wesley.
Tufte, Edward R.
    1990 Envisioning Information. Cheshire, CT: Graphics Press.
Vertelney, Laurie, Michael Arent, and Henry Lieberman
    1990 Two Disciplines in Search of an Interface. *In* The Art of Human-Computer Interface Design. Brenda Laurel, ed. Pp. 45–55. New York: Addison Wesley.

Technology for Failure: Skeptical Perspectives on Alternate and Hi-tech Teaching Methodologies

Gregory F. Truex

The invited session, "Closing the Educational Access Gap: Alternate and Hi-tech Teaching Methodologies and Their Implications," at the 89th Annual Meeting of the American Anthropological Association, brought together a group of scholars interested in the interface between the hardware of new communication technologies and users that engage these technologies in classrooms, laboratories, and remote computer conferences. The set of papers that composed this session can be divided roughly into those papers that focus on the teaching of anthropology and those papers that deal with the political or policy implications of the use of technology in education. This division is rough, however, and obscures their fundamental unity of perspective.

All of the papers, for example, dealt *centrally* with overriding social and cultural constraints and definitions that define an important area of anthropological interest and research. In this commentary, I attempt to draw out these anthropological issues within the context of the concrete examples presented by the participants. These issues, as I see them, are not the only ones raised by these papers. Other participants would, no doubt, emphasize other points implicit in the rich diversity of these presentations. In order to facilitate my synthesis, I will summarize briefly some of the papers' main points, recognizing that I have consciously and unconsciously been highly selective and that these summaries in no way substitute for a careful examination of the papers themselves.

Technology in Education: Skeptical Views

In their paper, "Interactive Courseware in Anthropology Classrooms," James F. Hamill and Linda F. Marchant reported on a project that used software that was designed specifically for anthropology courses. They use their experience to address two fundamental questions: (1) Can computer technology enter the lowest level of college teaching (effectively)? and (2) Is it worth the effort, and do students learn better? Hamill and Marchant are concerned in specific terms with bringing experimental and exploratory modes of learning to anthropology courses, supplementing or supplanting the symbolic learning modes of the traditional classroom.

Based on their experiences, they evaluate available anthropology teaching software on two dimensions: the first dimension (*venue*) examines whether the courseware is most appropriate for in-class or out-of-class assignments, and the second dimension (*range*) considers whether the courseware is related only to specific course content or if it can be related to a wide variety of teaching goals. This classification scheme is certainly general and can be applied usefully to much courseware, in addition to those reviewed by Hamill and Marchant, and is a nice contribution in itself.

Hamill and Marchant also provide a description of some very practical steps to take in implementing these technological innovations in the classroom. Providing detailed, step-by-step, keystroke-by-keystroke instructions, along with program demonstrations, seems essential for success. They address in practical terms how some of the interfacing problems addressed in other papers might be ameliorated or overcome. They also point out that there are costs: high-tech equipment may have to be moved with low-back labor.

The published version of Mason Weiss, Metzger, and McDonald's contribution cannot do justice to the actual, real-time presentation made during the session. Paradoxically, the most abstract paper was presented in the most concrete (and high-tech) way. In their paper, "Hypertext Indexing Applied to Computer Mediated Conferencing and Teaching: An Aid to Group Memory," Mason Weiss, Metzger, and McDonald are concerned with the eminently ethnological task of introducing organization to the notes and replies of a computer-network conferencing system. These notes and replies constitute the deliberations of the group participating in the conference. Because the conference is only minimally sequential—that is, "replies" *may, but do not have to, follow* "notes" that introduce a topic for deliberation—the logged record of the participants' contributions is in some sense linear, but not necessarily temporally or conceptually sequential. For this reason, the content of even a short conference quickly passes from the individual and collective memory of its participants.

To make this content accessible for the work of ongoing conferences—that is, to enhance the group memory—Mason Weiss, Metzger, and McDonald propose the use of HyperRez, Neil Larsen's trademarked Hypertext software. They would disentangle the strands of cohering topics from the temporal ordering imposed by the conference's log while retaining the ability to make or point to all available deductions implicit in the complete conference representation. This is achieved by creating a network (rather than linear) order of the notes and responses.

Mason Weiss, Metzger, and McDonald are able to give conference participants access to the "knowledge" being produced by their practices. The cognitive residue (as they call it) of these conference practices represents an emergent culture characterized by the more or less coherent body of belief implicated in the notes and replies that are its attestations. Alas, just as in all ethnographic work, knowing the relations between the order given by the ethnologists (Mason Weiss, Metzger, and McDonald) and the order implicit in the "culture" of the participants is problematic. Systematic work might be done to clear up just what these relations might be.

Perhaps the most common "high-tech" teaching innovation is the telecourse. Edwin Segal, in "Distance Education in Anthropology: Telecourses as a Teaching Strategy," raises a number of the most provocative issues concerning these education-at-a-distance mechanisms. For example, the materials that are available in such telecourses are never completely adequate; they are biased and distorted and do not represent the instructor's point of view, because, as Segal says, they represent "someone else's agenda." Of course, this problem is common to the selection of any course materials, unless you write the text yourself.

Segal attempts to assess the educational impact of a telecourse by comparing it with an ordinary course offering. The design problems in such a comparison are difficult to overcome, and Segal's analysis is not altogether satisfying. There is no straightforward way to separate the potential effects of sampling in different student populations from the effects of different teaching media. Segal does gather a number of interesting statistics about the different groups. While they lack explanatory value, in the statistical sense, the background data on the students suggest avenues for more ethnographic exploration of the issues Segal wishes to address. A better, contextualized understanding of the differences in student populations might make a more efficient experimental design possible.

In his article "Interpenetration of Technology and Institution: An Assessment of an Educational Computer Conferencing System," James McDonald confronts the disheartening experiences of community college students who attempted to take their first computer-based courses. His critique points to the bureaucratic logic that determines the disposition of computer-use in such a way that it is unequally accessible by members of different social categories: some white males find it easy and efficient; overall, minorities and women do not. Unhappily, those "marginal" students who, on economic and social grounds, might gain the most from distributed processing of their education (i.e., being able to take courses off-campus at times that fit around work schedules) are least supported by the hardware and software configuration of the campus.

McDonald's analysis of why this comes about involves two fundamental aspects. On the one hand, the technology, itself, is a dominant factor in determining the educational outcomes. The technology "formats" the educational event, so that only those with compatible attributes fully benefit. Thus, some students, in this case the less sophisticated users, are unable to access the technological benefits that are readily available to home-based computer owners. On the other hand, the bureaucracy's definition of the desired outcome, "computer literacy," guarantees that student experiences will not broaden their perspective and, thus, they will not *learn how to learn* beyond the merely technical application at hand.

Gail Bader and James M. Nyce, in "When Freedom of Choice Fails: Ideology and Action in a Secondary School Hypermedia Project," present an interesting vignette of a high school hypermedia course on American history and literature. The hypermedia projects described by Bader and Nyce reveal that from the teacher's point of view, the nonlinear, exploratory styles of learning promoted by hypermedia advocates are successful only if they lead students to find the "right" connections—in addition to connections leading from the students' own ideas and interests. To achieve this success, the teacher must structure the assignments and questions to assure that students will "discover" the correct relations. This structuring detracts from the element of "choice" that is embedded in hypermedia ideology.

Structure and Experience of Educational Technology

All new technology brings up questions of efficiency. This efficiency can be double edged. New efficient technology for education might be made cost-effective by displacing the jobs of anthropologists. In order to be efficient, in that sense, the technology must get the job done. Much of the evidence reviewed in these papers suggests that the job just does not get done. For one thing, the technology does not seem to serve the most appropriate constituencies. Paradoxically, equal access in absolute terms guarantees unequal distribution of benefits and inefficient use of resources. The failure rate in accessing education through high-tech is lamentably high.

All of the papers reflect a commonly held view that, while one educational goal of anthropology is to change attitudes, the social and educational constraints of teaching force us to focus on cognitive skills. Attempting to *measure* changes in student attitudes in our courses would almost be threatening to most of us: such intrusion on the affective person of the student violates our covenant to treat them as rational, thinking beings. This fundamental dichotomy in our approach to students can be seen by analysis of the technology/user interface presumed by the session's papers. The session's papers deal with three major dimensions of the technology/user interface: (1) locus of control; (2) quality of action; and (3) end goal definition.

Locus of Control

New technologies prompt our attention to very fundamental control issues that are integral to all social intercourse because new technologies require changes in existing behavioral norms. If nothing else, a new way of doing things (technology) changes who does what and with whom and redefines power and control. Many anthropologists seek new technologies for instruction in order to decenter and deconstruct the cultural assumptions of power and control implicit in traditional educational formats. Student choice and control in subject and approach is valued both for its humanistic and pedagogical wisdom. Implicitly at least, all of the papers agree that the effectiveness of teaching and learning is increased by active student participation.

Quality of Action

Following from basic cultural, and to some extent universal (cf. Osgood 1964), evaluative criteria, the session's papers reflect that active learning modes are preferable to (stronger than) passive modes. A sea of vacant-faced students, slouching inertly, is always a barometer of disengagement (often of the teacher.) Snappy rejoinders and lively debates are a sure sign of actual learning, in its fullest sense.

End Goal

The pedagogical hegemony of the acquisition of culturally defined (particularly Western) knowledge is no longer taken for granted. In particular, anthropologists are concerned, as are many academics, with students' experiential integration of their own alternative "knowledge."

The kinds of technological innovations on educational access that were considered during the session divide intriguingly along the dimensions outlined. For example, programmed textbooks and telecourses are most similar to the traditional classroom in terms of their locus of control (teacher structured), quality of action (passive, receptive), and end goal (imparting standardized symbol-dominated knowledge). Ethnographic films and videotapes are more like the "open classroom," where the learning of appropriate attitudes (however difficult these are to measure) and social skills are facilitated. These "open classroom" technologies are successful to the extent that they engage the student in active (quality of action) pursuit of their own *experiential* understandings (end goal). In general, however, these technologies are also structured (locus of control) from the outside and are presented in a set, predefined, linear order, just as the stations and opportunities in the "open classroom" are ultimately controlled by the teacher.

Hypertext allows significant degrees of student control but ultimately, as Bader and Nyce note, is structured by a teacher or compiler who is responsible in the current environment of education for imparting Western, symbol-dominated insights (i.e., knowledge as the end goal.) Also, hypertext is passive and receptive (quality of action) to the extent that students cannot create their own "structure"; they are allowed choices only within predetermined networks of connections (cf. Mason Weiss, Metzger, and McDonald, this volume).

In considering the various possible combinations of locus of control, quality of action and end goal, only computer programming itself and computer games have many of the qualities of student control, experiential end goals, and active quality. Both programming and computer games are similar to research in that their outcomes arise from the experience, itself. One does not emerge from a successful programming exercise or from winning a computer game with a "knowledge" of some set of actions. Rather, one has furthered the integration of skills with general applicability in the future. This integration is that of strategy—not how to solve the programming problem but how to *approach* the solving.

The real experts in computer games, 9- and 10-year-old children, have integrated the skills of anticipation and tactics. They have taken the "structure" of the game and learned to deal effectively with its various (perhaps infinite) permutations. Oddly, computer games, like the telephone, represent high-tech innovations that appear to present almost no class and gender barriers. Intriguingly, and significantly, I believe, that both computer games and the telephone share an "analog" input. If we use those technologies that are controlled with "joy sticks" rather than keyboards, we might get different educational results.

I do mean this facetiously, but only barely. The central issue is not the technology but rather the social definitions that envelope and determine both access and outcome of educational innovations. But these enveloping definitions are signaled by the input device: digital input implies, by and large, cognitive outcomes. It is no accident that the "user-friendly" interface of MacIntosh and "windowing" devices makes the computer more accessible to a wider range of users than the IBM/DOS command-line mode. Zapping a wastebasket is both symbolically and affectively satisfying. In fact, the difference between the expectations of the outcomes of a high-tech experience can be predicted, in this sense, rather well by knowledge of what the medium of input is. One can describe (deconstruct?) these as the relations *to* technology as well as, *sous rature, around* technology. In this, the session participants seem agreed: bridging the educational gap is not a hardware problem.

This is not to say that digital input is completely constrained. As I said, running throughout the papers, more explicitly in some, is a concern with the perceived disjuncture between changing the knowledge a student has (about anthropological issues) and changing the student's attitudes (about the subjects of anthropological concerns). The more valued outcome, attitude formation and change, is the raison d'être of anthropology but is viewed as utterly separate from the body of anthropological knowledge. Thus, one can teach and measure the students' knowledge of Nuer marriage customs but not their attitudes toward them. The assumption is that students are able to acquire cognitive knowledge without undergoing any particular affective change, but I wonder if this is necessarily true.

Hamill and Marchant detail uses of symbolically dominated courseware that would, it seems to me, have a decided "attitudinal" effect. Presenting the diversity of human cultural traits, using Douglas White's Cultural Diversity programs, makes the student necessarily consider each trait, however exotic it may be, in the context of other traits. The context (i.e., the range of related traits) explains the exotic trait under consideration. This drastically replaces the notion that an exotic trait is meaningful only if it is translated into our cultural symbol system with the notion that it is explained (predicted) by a range of universally associated traits. While this anti-ethnocentrism may not satisfy the most radical of interpretive, cultural relativists, it is a sure step forward from commonplace redneck bias.

Nonetheless, the demands of technology, computer technology in particular, imply very basic social interaction styles that may be rarely experienced by some users. The metalearning, or as Bateson calls it, deutero-learning, demands of the technology may be unfamiliar or even inappropriate for some users. McDonald noted the effects of technology on college students' communication styles and interactions. The fundamental aridity of feedback from high-tech mechanisms of learning are unlike any of the social interaction styles children are likely to encounter during normal enculturation. Thus, the fundamental demand characteristics of the technology discard the child's *deutero-learning* at the outset.

However, this may not have to be. Reduction of the social isolation of high-tech learning devices is feasible. Instead of parallel processors, parallel input devices, that access the same program and screen would allow students to learn through groups (cf. Stasson et al. 1991). Technologically mediated, experiential learning might be made more like the playground, where children learn about issues of "central importance to a democratic society—the interlock of order and flexibility, individual freedom and group consensus, stability and change" (Hawes 1974:21). Learning how to negotiate interpretations, within the bounds of symbol-dominated knowledge might be quite valuable. Such learning in a multicultural society might be imperative.

Anthropology seems to me not well practiced in dealing with local and emergent cultures. The Romney, Weller, and Batchelder (1986) Cultural Consensus Model frees the study of emergent culture from the bonds of tradition and provides a possible platform from which to measure the ongoing making of new cultures. Beyond fixed knowledge dominated by the symbols of tradition, culture also implies mutual agreement and reinforcement about how to proceed in solving problems. Culture is about both stability and change, about both law and order, as Hawes so nicely puts it, and achieving that order is a matter of metaconsensus. The agreement is not about what the order is but, rather, about what that order can be.

References Cited

Hawes, Bess Lomax
    1974   Law and Order on the Playground. *In* Games in Education and Development. Loyda
        M. Shears and Eli M. Bower, eds. Pp. 12–22. Springfield, IL: Charles C. Thomas Publisher.
Osgood, Charles
    1964   Semantic Differential Techniques in the Comparative Study of Cultures. American
        Anthropologist Special Issue 66(3):171–200.

Romney, A. Kimball, Susan C. Weller, and William H. Batchelder
   1986   Culture as Consensus: A Theory of Culture and Informant Accuracy. American
      Anthropologist 99(2):313–338.
Stasson, Mark F., Tatsukya Kameoa, Suki K. Zimmerman, and James H. Davis
   1991   Effects of Assigned Group Consensus Requirement on Group Problem Solving and
      Group Members' Learning. Social Psychology Quarterly 54(1):25–35.

About the Contributors

Gail Bader is an assistant professor (research) in the Department of Anthropology at Brown University. Her interests include technology and work in secondary and postsecondary education. She is especially interested in how American culture emerges and is reflected in all three of these areas.

Frank Dubinskas is the Howard W. Alkire Chair in International Business and Economics and associate professor of anthropology at Hamline University, which he joined in September 1993. His research and writing include a decade of work on the interface of technology, organizations, and culture, both in the United States and in Europe and Japan. Much of this work has been conducted in high-technology firms, and he has written on product development and manufacturing in the automobile, biotechnology, and computer industries; on concurrent engineering in manufacturing automation; and on knowledge management and collaboration in complex organization. His work has focused on the interactions among users, organizations, and flexible technologies, including electronic conferencing and other flexible software-based systems. Publications include the volume *Making Time: Ethnographies of High-Technology Organizations* (Temple University Press, 1988), as well as articles and case studies in management and anthropology. Before joining Hamline University, Frank Dubinskas was assistant professor in the Carroll School of Management at Boston College (1987–92); associate in research at the Harvard Business School (1985–87); and Exxon Fellow and visiting scholar in the Science, Technology, and Society Program at MIT (1983–85). During 1991 and 1992, he was NEH Resident Fellow at the School of American Research in Santa Fe, New Mexico, working on research materials from his (1990) field study of a manufacturing automation project at Apple Computer, Inc. Since 1988, he served on an AAA Panel on Disorders of Industrial Society, which will publish its work in *Diagnosing America* (University of Michigan Press, forthcoming).

James F. Hamill is associate professor of anthropology at Miami University in Oxford, Ohio. Since taking his doctorate in 1974 from the University of Wisconsin-Milwaukee, he has concentrated his research efforts on the relationships between culture, reasoning, and language, and has published several papers and a book on the subject. For the past five years, he has applied this interest to the area of personal computers as an instructional enhancement and is now developing the concept of "home-made software" for classroom and assignment uses.

Anna Hargreaves has been an instructional designer in the computer industry for ten years, most recently at Lotus Development Corporation, where she developed training courses for products such as 1-2-3 (spreadsheet) and Notes (group communications tool). Before her computer industry career, Ms. Hargreaves taught English and linguistics at Lanchester Polytechnic and Buckingham University in England and at Zagreb University in Croatia.

James H. McDonald is an assistant professor of anthropology at the University of Michigan-Flint. While pursuing a doctoral degree at Arizona State University, Dr. McDonald taught on and coordinated a local community college's computer conferencing system. In addition to his interest in the political and policy implications of technology, Dr. McDonald also has conducted research focusing on the Mesoamerican political economy. Recent articles include, "Whose History?, Whose Voice?: Myth and Resistance in the Rise of the New Left in Mexico" *Cultural Anthropology* 8(1), 1993; "Popular Justice in Revolutionary Nicaragua" *Social & Legal Studies* 1(2), 1992 (co-authored with Marjorie S. Zatz); and "Small-Scale Irrigation and the Emergence of Inequality Among Farmers in Central Mexico" *Research in Economic Anthropology* 13, 1991.

Linda F. Marchant is associate professor of anthropology at Miami University in Oxford, Ohio, where she has been a member of the faculty since 1989. Marchant's primary research area is behavioral primatology, with a special emphasis on lateral function. She recently completed a study of hand preference and tool use among the chimpanzees of Gombe. Marchant has published in the *American Journal of Primatology, The International Journal of Primatology, Journal of Human Evolution, Current Anthropology,* and *Behavioral and Brain Sciences.* She is also interested in models of human evolution and has a chapter in the book *The Archaeology of Gender.* As part of her interest in innovative teaching methods, Marchant has developed a course on primate behavior at the Cincinnati Zoo, and she and her colleague James F. Hamill are exploring the use of computers in teaching anthropology (this volume).

Audrey Mason Weiss is a doctoral candidate in anthropology at the University of California, Irvine. Ms. Mason Weiss has worked extensively with the BESTNet teleconferencing system and has research interests in northern Mexico.

Duane G. Metzger is professor of anthropology at the University of California, Irvine. Dr. Metzger has a long and distinguished research record in the area of belief systems and ethnosemantic analysis, set primarily in Mexico. In addition, he has spent a number of years developing the BESTNet teleconferencing system connecting students, faculty, and staff in institutions throughout the southwestern United States and Mexico in interactive computer conferences on diverse academic topics.

James M. Nyce is an assistant professor (research) in the Department of Anthropology at Brown University. Interested in how knowledge, technology, and innovation intersect and are historically constituted, he has studied knowledge representation and transfer in a number of academic, educational, and medical settings. He has published on work, technology, and knowledge construction in neurology, medicine, and higher education. With Paul Kahn, he has also edited *From Memex to Hypertext: Vannevar Bush and the Mind's Machine* (Academic Press, 1992).

Edwin S. Segal is professor of anthropology at the University of Louisville. His current research is concerned with images of race and ethnicity in South Africa. He has also done research on ethnicity and development in Malawi. In 1992, he was a Research Fellow at the Institute of Social and Economic Research, Rhodes University, Grahamstown, South Africa.

Gregory F. Truex is professor of anthropology at California State University Northridge. He received his M.A. at Tulane University and his Ph.D. at the University of California, Irvine. He also holds an M.B.A. from the University of California, Los Angeles. His principal fieldwork has been in Oaxaca, Mexico, where he began working 26 years ago. He is currently working on network approaches to rural Mexican social organization. In 1992, he ended a term as editor of *World Cultures Electronic Journal of Cross-Cultural and Comparative Research.*

napa bulletins

Why are anthropologists joining together in local practitioner organizations? What do anthropologists in government agencies do? How does one set up and operate a research and consulting business?

These are some of the questions answered in recent issues of the NAPA Bulletin, a monograph series for practitioners in the social sciences published semiannually by the National Association for the Practice of Anthropology, a Unit of the American Anthropological Association.

The following issues are now available:

2 Business and Industrial Anthropology: An Overview
Marietta L. Baba
$4.50 (members), $6.00 (nonmembers)

4 Research and Consulting as a Business
Nancy Yaw Davis, Roger P. McConochie, and David R. Stevenson
$2.00 (members), $4.00 (nonmembers)

5 Mainstreaming Anthropology: Experiences in Government Employment
Karen J. Hanson, ed., John J. Conway, Jack Alexander, and H. Max Drake
$2.00 (members), $4.00 (nonmembers)

6 Bridges for Changing Times: Local Practitioner Organizations in American Anthropology
Linda A. Bennett
$2.00 (members), $4.00 (nonmembers)

7 Applied Anthropology and Public Servant: The Life and Work of Philleo Nash
Ruth H. Landman and Katherine Spencer Halpern, eds.
$2.00 (members), $4.00 (nonmembers)

8 Negotiating Ethnicity: The Impact of Anthropological Theory and Practice
Susan Emley Keefe, ed.
$2.00 (members), $4.00 (nonmembers)

9 Anthropology and Management Consulting: Forging a New Alliance
Maureen J. Giovannini and Lynne M. H. Rosansky
$6.00 (members), $7.50 (nonmembers)

10 Soundings: Rapid and Reliable Research Methods for Practicing Anthropologists
John van Willigen and Timothy L. Finan, eds.
$10.00 (members), $13.50 (nonmembers)

11 Double Vision: Anthropologists at Law
Randy Frances Kandel, ed.
$10.00 (members), $13.50 (nonmembers)

12 Electronic Technologies and Instruction: Tools, Users, and Power
Frank A. Dubinskas and James H. McDonald, eds.

Please include payment, in U.S. funds, with all orders.

American Anthropological Association
4350 North Fairfax Drive, Suite 640
Arlington, VA 22203

Make your job easier with the

NAPA Directory of Practicing Anthropologists

Completely revised and expanded in 1991, the NAPA Directory gives you the detailed membership information you need to reach the growing sector of practicing and applied anthropologists.

Use the NAPA Directory to:

- Find hard-to-locate colleagues who are consultants, state or federal employees, or who work for private nonprofit organizations

- Locate practicing and applied anthropologists by area of expertise

- Obtain complete member information, including job title, affiliation, and specialties

Quantities are limited, so order now!

$2.50 (members) $4.00 (nonmembers)

Please enclose payment, in U.S. funds, with all orders.

American Anthropological Association
4350 North Fairfax Drive, Suite 640
Arlington, VA 22203

Double Vision: Anthropologists at Law

edited by
Randy Frances Kandel

As a result of increasing cultural pluralism and ever-sharpening economic disparities, the role of anthropological expert testimony is expanding beyond issue-oriented litigation about indigenous and minority peoples into more commonplace family, commercial, and criminal cases. *Double Vision,* the newest in the NAPA Bulletin series, is a reflective yet practical guide for anthropologists working or thinking of working with lawyers and for lawyers working or thinking of working with anthropologists. Written from the anthropologists' perspective, the book renders familiar the exotic culture of law, but attorneys may, with equal profit and pleasure, read it to gain insight into the exotic culture of anthropology.

Topics include:

☐ Differences between anthropological and legal methods

☐ Case studies of attorney-anthropologist collaboration

☐ Guidelines for the expert anthropologist

Members $10.00   Nonmembers $13.50

To order, write to:
American Anthropological Association
4350 North Fairfax Drive, Suite 640
Arlington, VA 22203

Please enclose payment with all orders.